Agrarian Structure and Peasant Politics in Scandinavia

SCANDINAVIAN UNIVERSITY BOOKS
Universitetsforlaget Oslo, Bergen, Tromsø
Munksgaard Copenhagen
Esselte Studium Stockholm, Göteborg, Lund

Agrarian Structure and Peasant Politics in Scandinavia

A Comparative Study of Rural Response to Economic Change

BY ØYVIND ØSTERUD

SCANDINAVIAN · UNIVERSITY · BOOKS

Universitetsforlaget
Oslo – Bergen – Tromsø

Section: B.40.29. – 012T

ISBN 82-00-01702-8

Cover design by Oddvar Wold

Printed in Norway by
Hestholms Boktrykkeri A.s

Distribution offices:
NORWAY
Universitetsforlaget
Box 2977, Tøyen
Oslo 6

UNITED KINGDOM
Global Book Resources Ltd
37, Queens Street
Henley-on-Thames
Oxon RG9 1AJ

UNITED STATES AND CANADA
Columbia University Press
136 South Broadway
Irvington-on-Hudson
New York 10533

To
Kirsten

Preface

The purpose of this book is to contribute to the comparative sociological study of peasant politics, which will be analysed as a response to the economic transformation of peasant society. The development of peasant politics is traced within the different agrarian structures of the Scandinavian countries from the late 18th century to about 1870.

The problem is stated in the Introduction, together with a short methodological and semantic discussion. The book is then divided into three parts, with nine chapters and a conclusion. Part I gives a theoretical discussion of previous works on peasantry and peasant politics, and concentrates on the conceptual and analytical debate on peasantry and the structural conditions for various types of peasant politics. Part II is a comparative analysis of Scandinavian agrarian structures previous to and during the great transformation of Nordic agriculture, while Part III first focuses on the concomitant development of political frameworks for peasant action, and then analyses the contrasting features of the peasant movement in each country successively. The basic configurations are connected to the transitional features surveyed in Part II, and the main lines of argument are summarized in a short concluding chapter.

Acknowledgements

I should like to record my gratitude to a number of institutions and people who have helped me with this work. The British Council supported me financially during my first session at the London School of Economics in 1971–72, and so did the Norwegian Research Council for Science and the Humanities during my stay in Norway and subsequently. The University of Oslo granted me generous leave while I was employed there during the later phase of this work, to fulfil my residence requirement in London. The Library Staffs at the L.S.E. and the University Library in Oslo have also been very helpful.

I am also grateful to several colleagues at the University of Oslo, who made valuable suggestions in the numerous conversations I had with them at an early stage. Dr. W. M. Lafferty read the first parts in manuscript, and his careful reading and illuminating criticisms were very useful. I also record my special thanks to Professor Ernest Gellner. His detailed advice on the general line of argument and on matters of style and presentation has been extremely valuable; and so have the encouragement and suggestions made by Dr. George Dawson.

Since a thesis essentially identical with the present work was awarded a University of London doctorate in 1974, I have nourished half-hearted plans to revise the manuscript thoroughly. I finally realized that for this type of work the need for supplements and revision would probably be endless. So I let it go with a number of minor changes only. I am particularly indebted to Professors Stein Tveite and Francesco Kjellberg for detailed comments. I have also received valuable suggestions from Ulf Torgersen, Stein Rokkan, Knut Dahl Jacobsen, Aleksander Loit, Torkel Jansson, Ingrid Åberg, and others.

I am afraid my shortcomings do not do justice to the advice received, so the responsibility for the final product is mine alone.

Contents

Introduction

On the problem

Until quite recently most major works on peasantry generally included some standard opening remarks expressing regret as to the paucity of such studies compared to the immense contemporary importance of the subject. This has changed lately. Peasant studies have become somewhat *in vogue* for anthropologists, sociologists, historians, and political scientists alike. Quantitatively, and maybe even qualitatively, the knowledge available to us is far better developed than a few years ago owing to a marked increase in valuable monographs and readings.[1] But there are still wide gaps, and the complementary interaction between the various academic disciplines leaves much to be desired. Peasants, as a specific socio-economic category, belong to nobody. I will not be concerned, therefore, with traditional boundaries in social science in this work, functional though they might often be. The core of the problem can be associated with what is usually called 'political sociology', although I try to make it a cross-disciplinary exercise, steered by the problem itself and by the subject matter rather than by artificial academic divisions.

Peasants should not be dealt with, as many high-level generalizations tend to do, in a spirit of global uniformity. Individual differences and geographical specificity may be just as important as the common fate of tilling the soil. Peasants and agricultural conditions of northern Europe, *in casu* Scandinavia, have been outside the mainstream of modern peasant studies. They are the subject of the present study–not in their contemporary condition (where the term 'peasant' on the whole seems to be a bit out of place), but during the last century's economic and political transformation of the Nordic countryside. The study covers the period from the late 18th century to about 1870, although some retrospective glances will occasionally be made at even earlier lines of development. In particular, I will stress the Norwegian case,

[1] Cf., *inter alia,* the bibliographical surveys in B. J. Siegel (ed.), *Biennial Review of Anthropology,* Stanford 1961, 1963, 1965 and 1967. Cf. also T. Shanin (ed.), *Peasants and Peasant Societies,* Penguin 1971; and J. M. Potter, M. M. Diaz and G. M. Foster (eds.). *Peasant Societies: A Reader,* Little, Brown 1967. See furthermore the references to Part I below.

but relate it systematically to a comparative investigation of the development in Sweden and Denmark. Although the political systems of the Scandinavian countries have become rather similar to each other, their historical and agrarian preconditions were remarkably different.

Generally, I shall endeavour to explain some of the socio-political implications of comprehensive innovations in peasant societies. I shall concentrate on the fundamental economic change implicit in the transition from subsistence to commercial farming, focusing on different types of political response to such change and relating them in turn to different characteristics of agrarian structure. The peasant reaction to the challenge of commercial agriculture will not be restricted to intensive studies of small-scale rural communities, but preferably analysed at the macro level, through its articulation on the national political scene.

The relevance of the topic, for both science and policy formation, should not require much argument. I hope it may give some useful insight into the background conditions for the development of industrial societies in the North or elsewhere. It may also, I suppose, be conducive to explaining policy formation and political alignments of lasting and present-day importance.

It should be stressed, then, that the aim of this work is not that of agrarian historiography. The history of peasant society, and of peasant politics, has been quite popular among Norwegian historians, and many outstanding studies of high quality have been published during the last century. I shall, of course, draw heavily on these works, but the approach, the context of intepretation, and some of the material of the present study will differ from that tradition. The task will, I hope, be supplementary rather than either redundant or antagonistic.

This work is, moreover, based on extensive utilization of (more or less) reliable secondary sources. No lengthy argument need be put forward in defence of this. The results of meticulous historical research are increasingly valuable for cumulative knowledge in so far as they provide–and are utilized as –a firm basis for synthetical accounts. The merits of a dialogue between description and theoretization, between accounts based on minute scrutiny of primary sources and more generalized synthetical efforts, should rather be a commonplace within the academic community.

Some reasons for the delimitation of regional area are perhaps necessary. Natural conditions and a common history have made Scandinavia a rather closely interrelated unit on the north-western corner of mainland Europe. All through modern history Norway, Denmark, and Sweden have formed varying patterns of interrelationships and political unions, adopting the characteristics of a regional system with important cultural, linguistic, and institutional similarities. The agricultural population in the three countries thus has relatively homogeneous characteristics– in spite of vast ecological variations–and this makes comparative investigation immediately more meaningful. The Scandinavian peasantry also developed its distinctive politics at the national level, a coherent articulation as a class 'for itself', within some crucial decades of the 19th century. At the same time, however, some striking contrasts in agricultural development mark the three countries off from each other. First, the time phasing and sequence of rural transformation through mercantilization of produce and individualization of tenure were quite different. In Norway the real breakthrough towards 'modern'–or 'neo-technical'–agriculture did not occur until the latter half of the 19th century, while in Sweden, and especially in Denmark, the 'agricultural revolution' began nearly a hundred years earlier.

Secondly, the structural differences were quite marked. In Norway the proportion of peasant freehold was early predominant, while the presence of a rural nobility was rather negligible. In Sweden the peasantry were dominated by a privileged nobility, which also formed the backbone of the state bureaucracy, and in Denmark the landlords and great landowners dominated the rural scene from their lowland estates. These preliminary remarks indicate some of the basic factors which are to be looked for, and which should be spelled out and qualified in the analysis which follows.

On method

The study is intended mainly as a total-system comparison, at the level of the nation-state, based on historical material. The interconnections analysed will be supplemented by further evidence from regional comparisons, particularly in the case of Norway, but lack of sufficient information has made a systematic inter-nation analysis on this level rather difficult to accomplish. The research covers an extensive span of time–from about the end of the 18th century to well into the second half of the 19th century–and is thus not intended as a micro-study based on a neat analysis of primary source material. Nor is it a comprehensive comparative analysis of many national polities categorized on a few key 'variables'. The research strategy is situated somewhere between these ends of the scale. By being limited to three countries, the scope for generalization is obviously reduced, but by analysing a rather detailed and nuanced range of information over a substantial period of time, the degree of specificity on the sequence of development is increased.[2]

In the research sequence three factors, with different methodological

[2] Cf. also the discussion by W. M. Lafferty, *Economic Development and the Response of Labor in Scandinavia,* Oslo 1971, pp. 26–27.

problems, may be singled out; the *descriptive* presentation of historical development and variance, the *primary analysis* of the described reality, and the *interpretation* of historical change.[3]

1) The descriptive exploration involves two main types of sources:
   a) Extensive use is made of secondary material, historical research, and narrative of events compiled by scholars in the field. To some extent cross-checking is made for the more crucial evidence and sources (cf. notes and bibliography).
   b) Statistical information and primary sources are also explored for further evidence. It should be noted, however, that the statistical information available is generally better and more substantial for the later parts of the period. Some judgements as to the reliability of such information are made in the relevant footnotes and references.

2) Primary analysis of the described reality is intended to imply a comparative mapping of time sequences both as to socio-political combinations and to economic processes and events. Not only are there structural differences between the three countries, but structural configurations and patterns of development occur at different times in the three polities. Thus the status of social 'facts' has to be separated out in its own historical context, and social 'archaisms', relics of past historical problem situations, should be identified and dated.
3) The interpretation of historical change is a further refinement of the analysis indicated above. It means not only that the structural relationships and the overall directions of change–from subsistence production and community localism towards market integration and cohesive nation-building; from communal landholding towards individualized agrobusiness–should be accounted for, but that the specific complexity of processes of development should be figured out. There are regional and geographical variations during historical transformation, some areas and part-societies being, in conventional language, more 'backward' than others, or being on another path. There might also be an important co-existence of historical residues–archaic patterns of the past–which are crucial to historical change on several levels:

- ideologically as the cognitions, values, myths, or folklore of the hinterland
- structurally as the peasant family farm or the farm structure itself
- institutionally as legal arrangements, systems of inheritance, or obligations.

[3] See also the general methodological discussion by Henri Lefebvre, 'Perspectives de la sociologie rurale', *Cahiers internationaux de sociologie,* XIV, 1953, reprinted in Lefebvre, *Du rural à l'urbain,* ed. anthropos, Paris 1970.

The study is, as stated, based on statistical information and empirical data wherever possible within the scope of the problem. A rigorous quantitative test of the validity of the argument is hardly attainable, however. The nature of the problem does not seem to lend itself to the use of probability theory where the impact of different factors is measured as different percentages of the variance. Thus, when talking about causation, the relative weight of different factors will have to be discussed rather tentatively, with no concrete attempt at definitive statements. It should, of course, also be noted that statistics *always* need interpretation within the given historical context. The topic here is not the various and changing political response of uniform social units, but rather the evolution of different phenomena hidden by a stable *name;* i.e. the notion of a peasant–*bonde*–has different *meanings* according to the actual structural transformation of the social reality within which people work and live. The Nordic peasant during the period under consideration is thus no fixed character but a changing phenomenon. Therefore, a mere mechanical comparison of 'indicators' and 'time series' of factors attributive to this phenomenon is not really to the point. In accounting for qualitative change like the transition from one type of social organization to another there are, as Barrington Moore has pointed out, some limits to the profitable use of statistical procedures.[4] This does not mean, of course, that further insight into the mechanisms of change and the links between interrelationships is unattainable by further use of quantitative methods. But such a goal would require another research strategy: one which would try to follow up, by strict analysis of ecological units considerably smaller than the nation-state, more specific hypotheses formulated in the present study. In this respect–the concession should not be stretched too far–this might be identified as quite similar to the 'explorative studies' of conventional methodology textbooks.

Note on semantics

The main conceptual problems of the present study are connected to the notion of a 'peasantry' as distinct from other types of social entities. This being rather crucial to the analysis undertaken, substantial parts of Part I are devoted to a conceptual and theoretical discussion of 'peasants' and 'peasant societies'. We also face, however, a linguistic problem, which should be given some short consideration. The Norwegian word for a peasant, *bonde,* is not quite synomymous with the general connotation of the term.[5] A *bonde* includes what in English is usu-

[4] *Social Origins of Dictatorship and Democracy,* Beacon Press 1967, p. 519.

[5] The word *bonde* is used in Swedish and Danish as well, but the Danes also employ the term *landmann,* which might be somewhat closer to the English 'farmer'. The Norwegian term *gårdbruker* can definitely not be distinguished from *bonde* in this way.

ally called a 'peasant', but it is also used in a wider sense as applicable to a 'farmer', a 'tenant farmer', and virtually anybody who owns a farm or tills the soil for a living, excluding only domestic servants and landless farm labourers (who, however, are usually included in the Norwegian term for the peasant community, *bondesamfunnet)*. The conceptual distinctions that are made in English are then rather non-existent in Norwegian (and in Swedish and Danish as well). But the word *bonde* does not have, on the other hand, the somewhat modest and subordinate and dependent connotations of 'peasant', probably owing to a historical situation in which the dominant Norwegian agriculturalist was a freeholder without noble lords controlling him on the land; and a *bonde* is often–in later years–differentiated from a *småbruker,* a smallholder who usually has too little land to be able to make a living exclusively from farming. The implications of various semantic distinctions will be further spelled out in the relevant chapters.

Some remarks should also be made as to the notion of 'agrarian structure'. The concept is used here in a rather broad sense to mean a set of institutions and socio-economic relationships governing the use of land as a factor of production. It includes the narrower term 'land tenure system', but has a wider application, covering also those aspects of social hierarchy, authority relations, economic obligations, and legal rights which are connected to the system of access to land. This term involves no special linguistic difficulties as the Norwegian equivalent is often the direct translation *agrarstruktur,* or *jordbruksstruktur.*

The outline

Part I of the present study is a theoretical discussion of previous works on peasantry and peasant politics. Starting (chapter 1) with a conceptual review of various notions of 'peasants', we next (chapter 2) consider the now classical controversy over principles for analysing non-industrialized economies, spelling out its implications for the identification of peasant societies. Then follows (chapter 3) a discussion about 'peasants in time', their classification as social entities in relation to the static-dynamic dichotomy. The historical diversity of peasantry is then (chapter 4) considered in a review of relevant studies of the structural conditions for various types of peasant politics.

This is to give the actual research background for the comparative analysis of Nordic peasant societies.

Part II is a comparative analysis of Scandinavian agrarian structures previous to and during the great transformation of Nordic agriculture, which, roughly, began in Denmark in the latter half of the 18th century, followed by Sweden, and then by Norway three-quarters of a century later. Chapter 5 focuses on the 'old peasant society' in the North in an attempt to pin down the structural characteristics which were transformed by, and which also made their marks upon, the transition in the following decades. Chapters 6 and 7 focus on the mechanisms of transformation themselves; first the various phases in the change of land tenure, individualization and development of freehold, and then the process of commercialization as the fundamental breakthrough towards market integration and 'neo-technical' agriculture.

Part III presents an analysis of peasant politics concomitant to agrarian change. The integration and articulation of peasantry within the different national polities are considered in chapter 8, while chapter 9 focuses on the comparative features of the 'peasant movement' in relation to the sequences of change. The general strategies and alliance configurations are compared in a final section.

The *conclusion* gives a short recapitulation of the major lines of argument related to the theoretical discussion in Part I.

It should be noted, finally, that particular facts or events are occasionally presented in more than one chapter. This is motivated by different contextual perspectives.

PART I:

Peasantry and Peasant Politics

The intention of Part I is to set the stage, in conceptual and theoretical terms, for the empirical inquiry in the later parts. The first chapter is a conceptual discussion of peasantry *sui generis,* and a review of various theoretical approaches to peasantry as a specific social type. Chapters two to four discuss analytical contributions to the study of peasantry *in context:* first the controversy over the analysis of peasant *economy,* where the economic specificity of peasant society is related to the conceptual variety dealt with in the first chapter. Then, in the third chapter, the notion of peasantry is explicitly discussed in its *historical* dimension, on the axis between static and dynamic phenomena. This is conceptually relevant so far as conceptual divisions may be broken down through social change, and a peasant may thus lose his intrinsic characteristics by the development of new types of social formations. The fourth chapter is an extensive survey of the *political* study of peasantry; how peasant politics might be explained, from different theoretical angles, in its relation to social structure and processes of economic change.

We are thus trying to relate the economic specificity of peasantry to peasantry as a historical phenomenon and to the political sociology of peasant behaviour. This selection of aspects is optional, and intends to construct a perspective on peasantry which will be employed in the empirical parts of the study, both as implicit background and as explicit points of reference.

CHAPTER 1

Peasantry–a conceptual delineation

The terms 'peasant', 'peasantry', and 'peasant community' have been employed to characterize such a wide variety of social groups that a consistent and non-controversial definition does not seem to exist. There has also been a tendency to dispute the very existence of peasantry as a valid concept, or to treat it as a rather bodyless notion. The existence of peasantry as a specific social structure can, however, be claimed according to the nature of our task. Empirically the Scandinavian agriculturalists of our present study have some general attributes in common with rural dwellers in areas as far removed in ecological and social conditions as Eastern Europe, Asia, Africa, and Latin America. But the conceptual and sociological generalization does not imply the existence of uniform and homogeneous social groups. Even if the social reality is rather unique, comparative investigations imply both differences and fundamental similarities, without which any general knowledge, and indeed any conceptualization, would fade away and dissolve. Thus, in order to understand any actual social reality we have analytically to abstract from it.[1]

The conceptual variety

The peasant studies of contemporary scholarship are influenced by different theoretical traditions and schools of thought. The various types of conceptual approach should be discerned, to sort out their implications for the identification of peasantry as a social entity. Four major traditions can be listed as follows:[2]

1) The ethnographic cultural tradition.
2) The dualist society thesis.
3) The specific economy approach.
4) The socio-economic class analysis.

[1] It is a methodological triviality that conceptual determination implies an abstracted selection of distinguishing characteristics. Teodor Shanin also employs this point in his introduction to the reading *Peasants and Peasant Societies*, Penguin 1971, p. 13.

[2] This selective division is highly stimulated by Shanin's typology in "Peasantry: delineation of a sociological concept and a field of study", *Archives Européennes de Sociologie*, Tome XII, no. 2, 1971, pp. 289–300.

It should be noted that elements from these conceptual traditions may be combined in the approach of individual writers, although their conceptual emphasis often stresses one particular tendency.

The ethnographic/cultural tradition tends to stress the traditional inertia of peasant quality as a left-over from an earlier national history, a 'cultural lag' which has to disappear before 'modernization' can take place. This tradition is interested in a particular 'peasant personality', where cultural traits tend to be defined in individualized terms. Shanin[3] names George Foster as a modern representative of this approach, but it should be added that Foster himself also points out his affiliation with the second tradition, saying that 'one of the most obvious distinctions between truly primitive societies and folk societies is that the latter, over hundreds of years, have had constant contact with the centers of intellectual thought and development . . .'.[4] The village character of peasant culture is confirmed, but seen as a local expression of a wider cultural environment.

The dualist society tradition explicitly stresses the partial nature of traditional peasant society in its relation to the modern urban centres. Peasantry is characterized as having a non-independent, symbiotic status, marginal to and yet dependent upon the world of cities. *Kroeber* stresses the fractional, incomplete quality of peasant society and culture, arguing that: 'Peasants are definitely rural–yet live in relation to market towns; they form a class segment of a larger population which usually contains urban centers, sometimes metropolitan capitals. They constitute part-societies with part-cultures.'[5] This is closely in accord with *Redfield's* condition for the usage of the term, when he stresses that the peasant community's system of values is 'consistent, in the main, with those of the city people who constitute, so to speak, its other dimension of existence'.[6] Redfield locates peasantry in an intermediate position between self-sufficient segments of the 'folk' societies and the modern societies of 'organic' interaction.[7] The cultural dimension of the issue is stressed when the 'little tradition' of peasantry is opposed to the 'great tradition' of the gentry.[8] *Fallers* is also trying to specify the part-society thesis in a clearer differentiation between peasant society and both the tribal and the modern

[3] Ibid.

[4] G. M. Foster, "What Is Folk Culture?", *American Anthropologist,* LV, no. 2 (April–June 1953), p. 164. In Foster's language 'folk' seems equivalent to 'peasant', even though Redfield among others distinguishes between the two.

[5] A. L. Kroeber, *Anthropology,* Harcourt Brace, New York 1948, p. 284.

[6] Redfield, *The Primitive World and its Transformation,* Cornell 1953, p. 40.

[7] 'Organic' interaction is here opposed to the 'mechanic' one in the meaning elaborated by Durkheim. Tønnies and Maine are other classical authors in the dualist society tradition. Cf. Shanin, "Peasantry: delineation . . .", op.cit.

[8] Cf. Redfield, *Peasant Society and Culture,* Chicago 1956.

industrial varieties, saying that: 'A peasant society is one whose primary constituent units are semi-autonomous local communities with semi-autonomous cultures.'[9] Peasantry is thus *by definition* related to a higher urban culture and a wider societal order.

The structural character of peasant communities as horizontal segments of a wider society is further specified by Julien *Steward,* followed by Eric *Wolf.*[10] Steward stresses the ecological and economic aspects of complex societies, which are split into part-societies composed of *'vertical segments'*, i.e. local units such as households, neighbourhoods, and communities, *'horizontal segments'* such as occupational, class, ethnic, or interest sub-societies which transcend localities, and finally *formal institutions* such as monetary system, education, church, official doctrine, which run through the whole large society. Wolf too is preoccupied with such network relations where peasantry depends on a rural-urban antinomy and inter-relationship. But he stresses that this interrelationship implies a power structure in which the peasant holds a subordinate position. Peasants are seen as 'rural cultivators whose surpluses are transferred to a dominant group of rulers that uses the surpluses both to underwrite its own standard of living and to distribute the remainder to groups in society that do not farm but must be fed for their specific goods and services in turn'.[11] Thus Wolf, as well as Steward, has rejected the cultural emphasis of the part-society thesis, and finds the crucial defining elements in the realms of land use, property holding, and trade. The basic peasant–nonpeasant contrast is not the axis between folk and sophisticate, but between occupational categories and between power groups–the agricultural producer in his corporate local setting, on the one hand, and businessmen, public officials, and lords of the land, on the other.[12] In these relationships peasantry is generally subject to structural subordinacy.

The common anthropological background of the dualist society tradition, as it originates from Durkheimian sociology, is mirrored in the separation of peasant society from tribal society. While tribal society is regarded as more complete in itself, peasantry is characterized by its partial nature. As for Wolf's concept of peasantry this is, as we have seen, just one aspect of the defining characteristics.

The specific economy approach is also a rather heterogeneous tra-

[9] Fallers 1961 (note 25 below), here in Dalton, ed. *Economic Development and Social Change,* Doubleday 1971, p. 171.

[10] See Steward, "Area research, theory and practice", *Social Science Research Bulletin,* No. 63, New York 1950; and Eric Wolf, "Types of Latin American Peasantry: A Preliminary Discussion", *American Anthropologist,* LVII, no. 3, 1955. See also Clifford Geertz, "Studies in Peasant Life: Community and Society", *Biennial Review of Anthropology,* 1961, pp. 1–41.

[11] Wolf, *Peasants,* 1966, p. 34.

[12] Cf. Geertz, op.cit.

dition. Generally the social structure of peasantry is being determined by the special economic operation of the family farm. This perspective is elaborated in *Chayanov's* theory of peasant economy, which will be treated in further detail in the next chapter. Raymond *Firth* also defines 'peasants' by economic criteria, meaning: 'A system of small producers, with a simple technology and equipment, often relying primarily for their subsistence on what they themselves produce.'[13] Firth also includes in the general category of 'peasants' not only landowners but tenants, peasant craftsmen, peasant marketers, part-time cultivators, and other countrymen, so far as they are socially integrated–through kinship and mutual interdependence–in the peasant community. Firth favours, in the case of low technology and subsistence production, 'the concept of 'peasant' . . . viewed as indicating a set of structural or social relationships rather than a technological category of persons engaged in the same employment'.[14] Thus he argues that the close corporate group characteristics of peasant communities could make rigid differentation extremely artificial and inappropriate. Firth is then not conceptually preoccupied with distinctions between peasant and tribal societies–the latter presumably characterized, according to 'part-society' theory, by a stronger self-sufficient identity. But the specification of peasantry as a non-industrial type, based on family labour, distinguishes the peasant from more modern variants. Again, according to Eric Wolf, the peasant, as a rural cultivator with certain rights in land, is different from a capitalist farmer, since his ultimate aim is subsistence and not reinvestment; 'he runs a household, not a business concern'.[15] The particular economic characteristics of the peasant family farm are crucial here. A general argument might be concentrated as follows: peasant economy on the level of the individual household unit has a specific dual character: the peasant family household is the unit of production, and the same household is the unit of consumption–i.e. the peasant farm is both an enterprise and a unit of domestic economy.[16] While the main aim of peasant economic activity is subsistence, it is simultaneously based on two different, and sometimes contradictory, principles. One is the long-term calculation of profitable reinvestment–which might be rather pronounced or quite negligible (but enough to reproduce the productive potential). The other is the permanent household requirements. The basic subsistence orientation of peasants does not imply production exclusively for their own immediate needs. Orientation towards production for the market does not only indicate the intensity of the element of enterprise, as

[13] Firth, *Malay Fishermen: Their Peasant Economy,* Routledge 1951, p. 84.

[14] Cf. Firth and Yamey, *Capital, Saving and Credit in Peasant Societies,* Lond. 1964, p. 18.

[15] Wolf 1966, op.cit., p. 2.

[16] This is a main point in Boguslaw Galeski's diagnosis of peasantry; cf. Galeski, *Basic Concepts of Rural Sociology,* Manchester University Press 1972.

Galeski argues (Galeski 1972:14), but also that market orientation may be a strategic option to uphold a precarious balance of livelihood in the interaction with natural, political, and economic environments.[17] Thus a double orientation towards both market and subsistence, rather than a specific 'subsistence economy', may be seen as the fundamental starting point.[18] Unfavourable product prices or new tax burdens (connected to monetary surplus) may then stimulate withdrawal from the market, back to substistence production, according to the strategic power and options available. It should be noted here that while the peasant farm, owing to its dual character, lacks the basic characteristics of a capitalist enterprise aiming at reinvestment for profit, changes in the mode of peasant farming are influenced by the functions and changes of the capitalist system as a whole. The overall growth of market relationships may in itself change the characteristics of peasant economy, and give it certain features specific to the capitalist enterprise.[19] Thus commercialization of agriculture implies, in the economic sphere, a change in the relative weight of the element of domestic economy and the element of enterprise; i. e. a growth of the entrepreneurial component and a weakening of the family character of farming. Conceptually, this might be labelled a 'de-peasantization' of the socio-economic entity. The transition towards the category of 'farmer' is thus a process of becoming, and the subsumption of a social group to any of the general types is more easily done the more clear-cut the group characteristics. By the relative growth of money exchange, commercial produce, and hired labour,[20] the peasant increasingly assumes the characteristics of the ideal type farmer–an agricultural business manager with farmland as one among alternative investment options.

Peasant economy is thus generally regarded as familistic and consumption-based, and its relatively autonomous character is supposed to disintegrate by widespread market integration and money exchange.

The class analysis of peasantry has traditionally been close to the Marxian approach. Marx was inclined to see peasantry as a left-over from an earlier social formation, as the suppressed and exploited producers of pre-capitalist society.[21] The idea of an asymmetrical power

[17] This is further elaborated by Eric Wolf, *Peasants,* Prentice-Hall 1966, pp. 16–17.

[18] Cf. Daniel Thorner, "Peasant Economy as a Category in Economic History" 1962, reprinted in Shanin, ed. *Peasants and Peasant Societies,* op.cit., pp. 202–218.

[19] Cf. Galeski 1972, op.cit., p. 22.

[20] S. H. Franklin says that "Capitalist farming is distinguished from peasant farming because primarily hired, as opposed to family labour, supplies the bulk of the manpower", in Franklin, *Rural Societies,* Macmillan, London 1971, p. 41 (note 38).

[21] Marx's rather fragmentary discussion of pre-capitalist economic systems is presented as a part of "Grundrisse der Kritik der Politischen Ökonomie" and called, in a separate English translation, "Pre-capitalist economic formations", Lawrence & Wishart, London 1964, with a useful introduction by Eric Hobsbawm.

relationship to the permanent disadvantage of the peasant is retained in some influential present-day formulations. Eric Wolf's political concept of peasantry implies, as we have seen above, a permanent and unfavourable charge on the peasant's production by dominant outsiders. And Daniel *Thorner* also stresses that while the main concern of the productive units in a peasant economy is to feed themselves, peasantry are also subjects exploited by others. In one way or another they must hand over, surrender, or sell to others part of their food crops, to sustain the State, the towns, or the local lords.[22] In contrast to primitive producers the peasantry thus produce for exchange, and 'the conditions of exchange are such that the peasants usually give more than they get'.[23] *Shanin* also approaches peasants in terms of power relationships, saying that they are rural cultivators with low technology who produce by means of family labour for their own consumption and 'for the fulfilment of obligations to the holders of political and economic power'.[24] A common denominator to these conceptual formulations is then that political dominance is included among the criteria presented to define a peasant. According to some scholars, this collective subordinacy within a wider order is the very core of the phenomenon.

These various conceptual traditions do not form watertight compartments. Contemporary scholarship is influenced by more than one of them, and elements from the originally more clear-cut trends are often combined to form an integrated construct. But examples of a more 'pure' conceptual approach do also exist. Foster might be said, although not without reservation, to belong mainly to the ethnographic cultural tradition, while Kroeber is close to the dualist Durkheimian one, and Firth to the specific economy approach–stressing the technological aspects. The adherents to class analysis originating from Marx are also numerous, although they often integrate elements from other trends into this analysis. Now, Redfield as well as Fallers may be said to combine the first and the second traditions, stressing the cultural dimension of the dualist 'part-society' idea.[25] Shanin and Galeski, on the other hand, present a combination of the third and fourth approaches–the specificity of peasant family farming analysed within the wider power structure which dominates peasantry as a class. The formulations of Wolf and Thorner seem to be the most embracing of those discussed

22 Thorner in Shanin, ed. 1971, op.cit., p. 206.

23 Thorner, op.cit.

24 Shanin, ed. 1971, p. 240.

25 Fallers argues for instance that African villagers should be called 'peasants' politically and economically (as semi-autonomous parts of a larger whole), but doubtfully so in cultural terms. See L. A. Fallers, "Are African Cultivators to be called 'Peasants'?", *Current Anthropology,* vol. 2, no. 2, April 1961.

here, since they add dualist society theory, the partial nature of peasantry, to the household economy as well as the politically subordinate character of peasantry. In their conceptual approach these writers thus share some characteristics with the influential school of Redfield and Kroeber, but turn another way as they specify the power relations between peasantry as an economic type and the non-peasant world.

'Peasant society'–the conceptual levels

It should be noted that the term 'peasant' has been employed to characterize different societal levels.[26] The conceptual reference may be features of *entire societies;* it may be the existence of *local communities* as parts of a larger whole; or it may be the character of *individual household units.* These different units in focus may be labelled the macro level, the medium level, and the micro level respectively. They are of course not totally separate empirical categories, but stand in a continuous functional interconnection.

At the micro level peasanthood is identified by the specific *internal organization* of the family household–its economic orientation and its principles of labour division and work. The subsistence orientation based on family work and non-industrial technology seems to be generally agreed upon.

At the medium level peasantry is founded on the particular *relationship* between the local community and a wider order. The fractional and semi-autonomous character of village life is stressed both by dual society conceptualization and by class society theory integrating the peasants in a nation-wide system of general stratification.

At the macro level a set of different criteria is combined as *aggregate characteristics* for determining whether the whole economy of a given country or nation should be regarded as a peasant economy. A relatively fruitful proposal could run as follows:[27]

- the society should be an agrarian society i.e. a substantial part of the population should be engaged in agriculture (Thorner's proposal is half or more of both population and labour force).
- the rural agriculturalists should be related, in cultural and economic terms, to a dominant urban minority (Thorner also includes strict political dominance–the conceptual fruitfulness of which will be discussed below).
- the territorial area should be politically integrated, as a country or nation, by a formal state power–to differentiate peasantry from the dominant clan or kinship integration of tribal societies.

[26] C. Daniel Thorner, "Peasantry", *International Encyclopedia of the Social Sciences,* Macmillan 1968, Vol. 11, p. 504.

[27] Cf. Thorner, "Peasant Economy . . .", op.cit.

– the socio-economic formation should be characterized by a typical unit of production – the peasant family household. Other economic units, such as capitalist farms (with hired labour) or plantations or haciendas may exist, but a substantial proportion of all crops should be grown by peasant households.

The criteria that are used here are then an aggregate of peasant characteristics at the medium and micro levels of social formation. The first criterion is a mere numerical requirement, while the second and the third specify the integration of part-society relationships, and the last one concerns the familistic organization of production units.

The simple features of peasantry

We shall now list some individual features which have been widely used in contemporary peasant studies to characterize peasants and peasant societies. The list will tentatively be organized from the most non-controversial down to the really controversial traits, and it presents rather concrete features. Their analytical status will vary according to the broader analytical framework (some traditions of which are outlined above), but they give some clues to the delimitation of peasantry as a general type of social formation.

1) Peasants are *rural cultivators* with *land husbandry* as the main means of livelihood.
2) The *familial household* is the basic socio-economic unit.
3) Non-commercial *subsistence production.*
4) A pre-industrial entity with *low technology.*
5) *Small communities.*
6) *Underdog* position.
7) *Managerial responsibility.*
8) Peasants are *proprietors* of land (?).

The first and second traits are made explicit in most definitions in the field. Land husbandry is also often seen to imply *settled* agriculturalists with a strong attachment to the soil. This means that nomadic pastoralists are excluded, but it usually includes swiddens, or 'slash-and-burn' cultivation.[28]

The third, fourth, and fifth traits seem to be less frequently pronounced, but nevertheless implicitly accepted by most scholars in the field. Firth includes the small community character of peasantry among his defining characteristics, as well as subsistence production at a low technological level.[29] Shanin, Wolf, and Franklin are all explicit about

[28] Cf. Wolf 1966, op.cit., pp. 21–25.

[29] Cf. Firth's introduction to Firth and Yamey, eds., *Capital, Saving and Credit in Peasant Societies,* Allen & Unwin 1964, p. 17.

the subsistence orientation, and Shanin also stresses the simple technical equipment in his definition of peasants.[30]

Points six and seven are of a more controversial type. In a strong interpretation of the underdog status of peasantry, favoured by Shanin, Wolf, and Thorner, as we have seen above, it is argued that a peasant is a rural producer who is exploited by someone else. This means a strong asymmetrical power relationship to the disadvantage of the peasant, implying an unfavourable charge on his production. This would probably exclude some rural freeholders, such as a greater proportion of the Norwegian ones of the present study, from the very definition of 'peasantry'. Ernest *Gellner* argues that it would exclude Berber cultivators of the Atlas as well, who are culturally integrated in a wider order although politically autonomous.[31] The concept of peasantry is made less empirically restrictive if the underdog position is given a weaker interpretation, implying only a generally low social status in a hierarchical economic and political order, whether the peasants are directly dominated by a class of landlords or not. But a landholding aristocracy is then in any case excluded from the notion of peasantry.

On the other hand, peasantry is often delimited downwards by the criterion of managerial responsibility, thereby excluding plantation workers and hired farm labourers.

The last feature, saying that peasants are owners of the land they work, is an even more restrictive requirement in the direction of formal control. But this criterion is highly controversial and in fact held by very few writers in the field.[32] It would exclude not only squatters but also tenants and sharecroppers from the category of peasants, and reserve the term for smaller groups of freeholders.

The legal status of peasants is generally regarded to be that of proprietors, tenants, or crop sharers, who may or may not be free to leave the land.[33]

Against the background outlined above, a general notion of peasantry might be delineated as follows:

From an economic point of view peasants are conceptually situated

[30] Shanin's definition runs as follows: "The peasantry consists of small agricultural producers who, with the help of simple equipment and the labour of their families, produce mainly for their own consumption and for the fulfilment of obligations to the holders of political and economic power", in "Peasantry as a Political Factor", Shanin, ed. 1971, op.cit., p. 240.

[31] E. Gellner, "The Great Patron. A reinterpretation of tribal rebellions", *Archives Européennes de Sociologie,* X, 1969, p. 67.

[32] Cf. Henry Landsberger, "The Role of Peasant Movements and Revolts in Development", in Landsberger, ed., *Latin American Peasant Movements, Cornell University* Press 1969, p. 3.

[33] Cf. Thorner 1968, op.cit., p. 504.

between 'farmers', on the one hand, and 'tribesmen', on the other. Socially their place is between the great rural aristocracy and the landless rural proletariat. But the more one specifies the characteristics of the pure types, the more infrequent they are as actually existent forms, and the more numerous are the intermediate and marginal cases. Borderline cases within the conceptual framework may be delineated as analytically marginal to the generic types, sharing some of the important characteristics with the hard core groups but not all of them, or only doubtfully so.[34] Such analytically marginal borderline cases may be remote mountain or frontier squatters, market-oriented or culturally integrated pastoralists, peasant-workers of semi-industrial kind, kibbutzniks, kolkhoz-members, tribesmen-becoming-peasants, or peasants-becoming-farmers (or the other way around). Conceptual marginality thus does not mean numerical insignificance.

But we will not make the core of the pure type peasants too restrictive and the range of conceptual marginality too broad. We will rather use the term 'peasants' in a fairly broad concept, closer to van Dietze's conceptual compromise: "Common to all the shifting meanings, however, is a view of the peasant as a tiller of the soil to whom the land which he and his family work offers both a home and a living."[35] We shall then specify our own usage of the term a short step further: The peasant is a rural cultivator based on family household economy and producing mainly for subsistence. He is related to a wider societal order, and generally has a low status within it.

Differences between types of peasantry as well as further qualities of peasant society as a social formation will be left for empirical discoverage.

The awkward class

The above indication of the 'low status' of peasantry within a hierarchical order is of course rather vague. We will, following *Shanin* again, try to localize the general status of peasantry in a wider society according to different concepts of social class.[36]

Following *Dahrendorf*'s criteria for definition of social class, according to which it is seen as conflict groups emerging from the distribution of power in society (Dahrendorf 1959), peasantry is likely to fall into the diffuse category of 'the ruled', together with other subject groups in society.

According to *Marxist* criteria, the relationship to the means of pro-

[34] Cf. Shanin, "Peasantry: delineation . . .", op.cit., and introduction to *Peasants and Peasant Societies,* op.cit.

[35] C. von Dietze, "Peasantry", *Encyclopedia of the Social Sciences,* 1934, p. 48.

[36] Cf. the discussion of this in Shanin, "Peasantry as a Political Factor", op.cit.

duction, peasantry would probably qualify for an anonymous place among the highly amorphous 'middle classes', or 'petite bourgeoisie'. Marx himself, writing about French peasantry in the middle of the 19th century, was not satisfied with this; 'In so far as millions of families live under economic conditions of existence that separate their mode of life, their interests and their culture from those of the other classes, and put them in hostile opposition to the latter, they form a class. In so far as there is merely a local interconnection among these smallholding peasants, and the identity of their interests begets no community, no national bond and no political organization among them, they do not form a class.'[37] Peasantry is then a partly inarticulate remnant of pre-capitalist society, a 'sack of potatoes' (Marx) not yet a class, but gradually becoming one.

Furthermore, for Max *Weber* 'class situation is . . . ultimately market situation'.[38] Common economic interest and destiny are conditioned by market relationships, with the themes of class domination or direct power less pronounced. To Weber social classes are differentiated according to specific kinds of property, trade and industry, or services offered in the market, opening up for an unlimited analytical division of society into small subgroups.[39] In this way, too, peasantry seems to fade away as a qualitatively distinct social entity.

To analyse peasantry as a social class, in any of the conceptual traditions outlined above, is then only partly to the point. Peasantry obviously have some common interests, sharing their ultimate life situation and culture in contrast to other parts of society. Under certain circumstances, when their situation is especially precarious or changing, they may cooperate in political 'class' action, as a common articulation against rural overlords, city bourgeoisie, or state power. But the structural characteristics of peasant society may be a countervailing force to such class-like articulation: the vertical segmentation favouring a corporate loyalty to household, homestead, or estate; the localistic orientation due to ecological situation; and the atomistic fragmentation due to isolating conditions of work and geographical setting. The class characteristics of peasantry are then a question of degree and of concrete historical situation, where even withdrawal from market relationships back to the closed corporate unity may be seen as a strategic reaction to pressure from outside.[40] We may at this stage quote from Shanin's article "Peasantry as a Political Factor" at some

[37] From "The 18th Brumaire of Louis Bonaparte", first published in 1852.

[38] From *Wirshaft und Gesellschaft* 1922, part III, shapter 4.

[39] Ibid. cf. Shanin, ed. 1971, pp. 252–253.

[40] Cf. Eric Wolf, "Types of Latin American peasantry: a preliminary discussion", *American Anthropologist,* vol. 57, June 1955, pp. 452–472.

length: 'The main duality in the peasants' position in society consists in their being, on the one hand, a social class (one of low 'classness' and on the whole dominated by other classes) and, on the other, 'a different world'—a highly self-sufficient 'society in itself', bearing the elements of a separate distinctive and closed pattern of social relations. The peasantry is the social phenomenon in which the Marxist approach to class analysis meets the main conceptual dichotomies of non-Marxist sociological thinking; Maine's brotherhood versus economic competition; de Coulangue's familistic versus individualistic; Tønnies' *Gemeinschaft* versus *Gesellschaft* or Durkheim's mechanic (segmentary) versus organic societies (Redfield 1955:139–143). This unique duality ('class' and 'society') leads to conceptual difficulties, yet may well serve as a qualitative definition of the peasantry—especially when differentiating this entity from wider, more amorphous groupings such as 'middle classes', 'exploited masses' or 'remnants of feudalism'.'[41] For the sake of empirical comprehensiveness, as stated earlier, we may make a reservation as to his remark about class-political subordination, but generelly this determination of kind seems highly fruitful. The vertical segmentation–the Dutch word *verzuiling* has become somewhat fashionable—into local communities is complementary to connections, of various strength, to the wider society as 'part-societies' (Kroeber, Redfield). Ecological isolation, backward technology, lack of experience, and, maybe more important still, divisions *within* these communities, may contribute to the relative low degree of 'classness' (subjective 'for itself' as well as objective 'in itself') and make the framing of national goals, symbols, leadership, and organization difficult. The structural and developmental prerequisites for such articulation, and the courses it takes, are then internal to the transformations of peasantry itself, as it is *defined* in its relations to the wider societal order.

[41] Shanin, ed. 1971, op.cit., p. 254.

CHAPTER 2

Analytical approaches to peasant economy

We shall now go further into the specific quality of peasantry by discussing how it might be analysed from an economic point of view. The various socio-political perspectives will be discussed in later chapters. We are interested in various conceptions of the *nature* of peasant economy rather than the technical aspects of agrarian economics. These analytical conceptions will illuminate the question of what is really generic about peasantry and peasant society.

The controversy over analytical approach in this field is a rather classical debate. We shall summarize the main arguments, trying to map their implications for the conceptual characterization of peasant economy. It should be noted that the *big debate* in question has not been restricted to the principles for analysing peasant economies, but covers the whole field of non-industrial and precapitalist societies, as studied in anthropology and economic history. The main topic of the debate is thus not the specificity of peasant economy as such, but how any economic formation prior to or outside the industrial market economy, peasant and tribal and 'ancient' alike, should be analysed. This debate, fundamental to the contemporary conception of social science, is then not actually concerned with the definition of peasant economy, but it might still illuminate how peasant economic behaviour, and the institutions within which that behaviour takes place, can be studied from different theoretical points of view.

The maximizing peasant

An influential trend has been labelled 'formalist', and traced back to the 'new economic anthropology' of Firth and of Herskovits.[1] This implies the application of an economic analysis originally derived from a capitalist market economy on peasant and primitive economies. 'Pre-

[1] Firth, *Primitive Polynesian Economy*, London, 1939, Herskovits, *Economic Anthropology*, New York 1952 (originally *The Economic Life of Primitive Peoples*, New York 1940); cf. E. E. LeClair and H. K. Schneider, eds. *Economic Anthropology: Readings in Theory and Analysis*, New York 1968.

modern' economies are seen as 'modified' capitalistic, with their functional equivalences of institutions like 'credit', 'saving', and 'capital' present in *forms* different from those in the capitalist commodity market with its general-purpose money as the standard of exchage. Different types of commodities, like salt, blankets, or cows, assume the form of special-purpose money–it's a difference of degree, not of principle. Thus the application of conventional economic concepts leads to a search for analogues, parallels, and similarities between economic institutions and mechanisms in the 'modern' world and the non-literate, primitive world.[2] The expressed theoretical prerequisite for this formalist orientation–the general applicability of conventional economic concepts–is the universality of decision-making behaviour. The relevance of formal economic theory to non-market situations is closely associated to Lionel Robbins' famous analytical conception of economics: 'Economics is the science which studies human behaviour as a relationship between ends and scarce means which have alternative uses.'[3] The principle of 'economizing' as the allocation of scarce resources among alternative ends is the fundamental premise of neoclassical, or marginalist, economic theory, where the calculus of maximization is supposed to be a general principle valid everywhere. When scarcity, choice, and maximizing behaviour are thus regarded as universal phenomena, maximization theory is universally applicable.[4] The underlying condition of the formalist trend is of course that the theory of maximization says nothing about *what* is maximized. The assumption that *profit,* in monetary terms, is the end maximized, is thus just an application of the theory, not the theory itself.[5] The principle is formal, implying that anything, prestige, status, power, well-being, money or whatsoever, may constitute the end. To specify the channelling of economic choice, it is often stressed how the decision-making process is circumscribed by the particular social structure,[6] or how social institutions themselves are generated by processes of strategic choice.[7]

[2] Cf. Firth and Yamey, eds. 1964, op.cit. An influential expression of this trend is also Sol Tax, *Penny Capitalism,* Washington 1953.

[3] Robbins' conceptual discussion was published in 1932, but is reprinted in LeClair & Schneider, op.cit., p. 96.

[4] Cf. R. Burling, "Maximization Theories and the Study of Economic Anthropology", *American Anthropologist,* 64 (1962), pp. 1179–1203.

[5] Cf. LeClair & Schneider, eds. 1968, introduction, p. 8.

[6] Cf. Manning Nash, *Primitive and Peasant Economic Systems,* Chandler publ. 1966.

[7] Cf. Fredrik Barth, *Models of Social Organization,* London 1966.

The Opposition: Different Modes of the Economy

The most consistent antagonist to the formalist trend may have been the 'substantivist' opposition of Karl Polanyi and his associates.[8] Polanyi questions the relevance of the maximization approach to non-market-oriented societies, and denies the general applicability of the logic of formal rationality. The focus is shifted from the general character of means-ends relationships to the specific variety of human institutions and economic categories, where many 'economic' non-western institutions are supposed to be quite distinct from the market system, thereby requiring alternative economic theories. Polanyi argues for a non-formal 'substantive' concept of the economy, and he defines this as 'an instituted process of interaction between man and his environment, which results in a continuous supply of want satisfying material means'.[9] On this basis Polanyi separates three different forms of integration of economies, as definite institutional arrangements within which individual behaviour takes place–such as the integrative pattern of *reciprocity* (a symmetrical relationship in economic transactions), the pattern of *redistribution* (with central points of authority), and the pattern of *exchange* (through an impersonal market system). Different types of economic systems are classified according to the dominant forms of integration, where 'the rise of the market to a ruling force in the economy can be traced by noting the extent to which land and food were mobilized through exchange, and labour was turned into a commodity free to be purchased in the market'.[10] It is further argued that the method and content of economic theory (i.e. marginalist analysis) were shaped by the growth of the market mechanism and the market principle integrating the components of the economy.[11] The market organization was said to compel the participants to seek material self-gain, as each had to sell something of market value to acquire the material means of existence.[12] In Dalton's words: 'Economic theory was created to analyze the special structures, processes, and problems of market-oriented industrialism, with its special features of all-purpose money, impersonal contract obligation, atomistic individualism, and the institutional necessity for individuals to acquire a livelihood through market sale of labour, natural resources, and out-

[8] Cf. Polanyi, *The Great Transformation*, New York 1944; Polanyi, Arensberg and Pearson, eds. *Trade and Market in the Early Empires*, Free Press 1957; George Dalton, "Economic Theory and Primitive Society", *American Anthropologist* 63 (1961), pp. 1–25.

[9] In Polanyi et al. 1957, op.cit., p. 248.

[10] Ibid., p. 255.

[11] Dalton, op.cit., p. 144 in LeClair & Schneider, eds., op.cit.

[12] Ibid., p. 145 ibid.

puts. It is these which create the gulf between the primitive and the West.'[13]

An important aspect of the argument that not all economies do operate according to the theory of the market model is Polanyi's distinction between economies which are 'embedded' and 'disembedded' in the social structure (Polanyi 1944). The organization of the market economy is seen as a 'disembedded' cohesive entity in itself, where economic processes can be analysed without reference to the social, since the market mechanism has removed the economic sphere from its integration with the rest of society. 'In contrast, the close integration with social and economic institutions in primitive society makes it impossible for the anthropologist to describe the economic, without at the same time showing its relation to the social.'[14]

It should further be noted, as the core of the substantivist position, that the formalist maximization analysis is generally regarded as a reflection of atomistic utilitarianism, in contrast to the analysis of various patterns of institutional integration. Polanyi himself stresses that the forms of integration are not merely reflected aggregates of the respective forms of individual behaviour, that 'if, in any given case, the societal effects of individual behavior depend on the presence of definite institutional conditions, these conditions do not for that reason result from the personal behavior in question'[15] This is, however, not only a proposition of empirical substance, but a general refutation of methodological individualism, which marks a further contrast to neoclassical economics.[16]

On this general level, though, it seems striking that the formalist-substantivist controversy does not in principle give different answers to the same questions, but the participants argue somewhat at cross-purposes, and ask different questions which, according to circumstances and subject matter, may be relevant in due course.[17] The history of the debate may, however, help to clarify the asking of questions. Some implications will be hinted at below.

[13] Ibid., p. 163, ibid.

[14] Ibid., p. 155, ibid.

[15] In Polanyi et al. 1957, op.cit., p. 251.

[16] The doctrine of methodological individualism, and the debate about it, has a very long ancestry in the history of the social sciences, cf. the 'reconsideration', with extensive bibliographical notes, by Steven Lukes in Emmet & Macintyre, eds. *Sociological Theory and Philosophical Analysis,* Macmillan, London 1970, Chapter 5.

[17] Cf. the compromise suggested by Frank Cancian, "Maximization as Norm, Strategy, and Theory: A Comment on Programmatic Statements in Economic Antrhopology", *American Anthropologist* 68 (April 1966), pp. 465–470.

The Counter-opposition: A Marxian Structuralist Synthesis?

The above tendency towards pluralistic harmony, embracingly tolerant as it is, cannot so easily swallow up a third, and more recent, trend in the theoretical discussion about peasant and pre-capitalist economies. A mainly French school of thought, including authors like M. Godelier, C. Meillassoux, and E. Terray, may be labelled 'Marxist substantivism', and seems highly influenced by the Althusserian variant of *structuralisme.*[18] This trend is in accordance with the substantivist critique of formal maximization theory. In Godelier's view the formal qualities of 'rational' economic behaviour are not sufficient to separate the particular economic sphere from other aspects of social activity. 'Economic rationality' has a genuinely different content in different types of societies, as for instance monetary maximization of profit is a characteristic of the industrial market economy. Specific types of societal structure thus provoke specific forms of rational behaviour. The apparently functional equivalences between elements in various economies are subordinated to the fundamental differences caused by the non-conforming social structures within which the element functions. However, Polanyi's diagnosis of the economy as 'disembedded' from the social structure of mercantile societies is not accepted by structural Marxism. There are no societies in which the internal relationship between the economy and the non-economic sphere is absent; their respective autonomy is only apparent–an untenable *erscheinung.*

The traditional substantivist analysis of the three principles of integration and distribution of resources–reciprocity, redistribution, and exchange–is also regarded as inadequate, since no criteria for evaluation of the distributed objects are implied. Generally the substantivist trend is criticized for focusing on the sphere of circulation and not on the relations of production, which is supposed to be the basic sphere of societal contradictions.[19] In contrast to the substantivist view of the 'embeddedness' of pre-capitalist economies, the economic base is given some universal priority as the decisive instance. It is argued, however,

[18] See Maurice Godelier, *Rationalité et Irrationalité in Economie,* Paris 1966; Claude Meillassoux, "Essai d'interprétation du phénomène économique dans les sociétés traditionelles d'autosubsistance", *Cahiers d'études africaines,* no.4, 1960, pp. 38–67, and Meillassoux, "From reproduction to production", *Economy and Society,* Vol. 1, no. 1, 1972, pp. 93–105; Emmanuel Terray, *Le matérialisme historique devant les sociétés segmentaires et lignagères* (à propos de *l'Anthropologie Economique de Gouro* de Claude Meillassoux), Maspero, Paris 1969. The most comprehensive selection of studies presenting the program of Marxian *structuralisme* is Louis Althusser et al., *Lire le Capital,* Maspero, Paris 1967; see also Althusser, *Pour Marx,* Maspero 1965. Especially Terray is programmatically close to the Althusserian school.

[19] Cf. Meillassoux 1972, op.cit.

that kinship relations may be the dominant forms of production in lineage and segmentarian societies, but that the role of the various levels in production is decided by the economic base. It is thus only under capitalism that this decisive base is also the dominant one.[20]

Conceptual implications of the big debate

Concluding the rough summary of the controversy so far, it seems likely that the formalist trend tends towards stressing the similarities between the economic category of peasantry and other types of economic formations. In a rigorous interpretation this implies definition of a 'peasant' as an economic phenomenon on a gradual scale from the pre-neolithic hunter to the white-collar entrepreneur; different in degree, but not in principle.[21] In a less rigorous one it implies no particular definition or notion of 'peasantry' whatsoever, but just states that any socio-economic phenomenon, of whatever qualitative characteristics, can fruitfully be studied within a general decision-making framework.

The substantivist position, focusing on other aspects of reality, stresses the discontinuity, the qualitative gap between institutions of uncomparable heterogeneity, within which the peasant would qualify as a rather unique and distinct entity.

Paradoxically enough structural Marxism, from this angle, seems to be in an intermediate position: the specificity of different types of economic formations is confirmed, but the continuity and relations between them are stressed through the focus on the economy as a generally decisive base, constituting an explanatory link of the transformation from one type of societal formation to another. The economic sphere, regardless of social formation or political superstructure, is then regarded as the 'privileged field' of social change.[22]

The theory of peasant economy

Three basic questions seem highly acute against the background of our discussion so far:

1) How is peasantry *as such* affected by historical evolution and mechanisms of social change? That is: should peasantry be treated as a dynamic and transitional phenomenon *sui generis,* or should the

[20] Cf. Terray 1969, op.cit., ch. 5; Godelier 1966, op.cit.; E. Balibar in Althusser et al. 1967, op.cit.

[21] But the particular characteristics of peasantry could of course be specified along other dimensions, marking it off from other social types. This seems most often to be done by social anthropologists, even if they accept the formalist analysis of economical aspects.

[22] An extensive interpretation of this position is given by Jon Elster, *Nytt perspektiv på økonomisk historie,* Pax, Oslo 1971.

character of actual dynamism be left for empirical discoverage and empirical generalization?
2) What is the distinctive peculiarity of peasant economy within the category of non-industrial economies?
3) To what extent are all principles of neoclassical marginalist analysis inevitably bound to a developed capitalist economy?

The first question, which also carries on our conceptual delineation in the first chapter, will be given a separate discussion in the next chapter.

From a certain angle the second question is closely affiliated to the last one. If marginalist principles are inapplicable outside a capitalist market econony, only two options are available. Either the analysis may still be applied to peasant economies, which are then considered capitalistic 'in principle', with their pre-modern functional equivalents for the institutions of a fully developed market economy. Or its relevance to the study of peasant economies may be denied, in favour of an alternative economic theory which can transcend marginalist principles, both to account for the specificity of peasant economic behaviour, and to explain the change from one dominant type of economic formation to another.

However, as clearly demonstrated by *Chayanov's* theory of peasant economy, the general principles of marginal utility analysis are applicable outside a capitalist market economy.[23] This contribution should therefore be shortly considered. Its answer to our second question above will thereby be clear as well. Chayanov concentrated his analysis on the peasant family farm, which employed no hired hands but relied on family labour only. On the basis of extensive *zemstvo* statistics, he formulated both a theory of peasant behaviour at the level of the individual farm, and a theory of peasant economy as a specific economic system. According to this theory, it was a misunderstanding to view the peasant family as a business concern, an enterprise of a capitalist sort; 'the structural peculiarities of the peasant family labour farm make it abandon the conduct dictated by the customary formula for capitalist profit calculation.'[24] The peasant motivation and behaviour, contrary to those of a capitalist, aim at securing the subsistence needs of the family household rather than at making a profit. The intensity of cultivation and the size of the net product is not determined by objective

[23] A. V. Chayanov was an influential heretic in the Russian debate on agrarian economics during the first three decades of this century. After the Revolution he became director of the Institute of Agricultural Economy, but was arrested in 1930 under the accusation of being a *pro-kulak* ideologist. See the English volume of *The Theory of Peasant Economy,* edited and introduced by Daniel Thorner, Basile Kerblay and R. E. F. Smith, Irwin, Illinois 1966.

[24] Chayanov 1966, op.cit., p. 7.

calculations of the highest possible net profit in the given market situation, but by the balance between subsistence needs and the subjective distaste of labour: 'the degree of self-exploitation is determined by a peculiar equilibrium between family demand satisfaction and the drudgery of labour itself.'[25] This 'labour-consumer balance' may be labelled the key to Chayanov's position.[26] The economic peculiarity of peasantry makes it a specific non-capitalist phenomenon, to which the classical and neo-classical economic theory of the capitalist entrepreneur does not apply. So long as the subsistence needs of the family are not satisfied, i.e. the labour-consumer balance is not reached, decreasing returns of the value of marginal labour do not hinder production; 'in conditions where capitalist farms would go bankrupt, peasant families could work longer hours, sell at lower prices, obtain no net surplus, and yet manage to carry on with their farming, year after year'.[27] This, in Chayanov's view, gave the peasant family farms an extraordinary competitive power versus large-scale capitalist farming, contrary to Marxist writings of his time. The important prerequisite for the non-usability of conventional economic analysis was the lack of fundamental factors required by the model, especially the lack of wages. The family labour product was effectively undifferentiated and indivisible, it could not be decomposed into wage and other factors; and 'since there is no social phenomenon of wages, the social phenomenon of net profit is also absent. Thus, it is impossible to apply the capitalist profit calculations'.[28] Capital investment is, furthermore, only advantageous to the family labour unit if it affords the possibility of a higher level of well-being, subjectively evaluated against the labour-consumer balance.[29] This balance is principally affected by differentiating demographical factors like the size of the family, the ratio of working to non-working family members, the 'natural history' (time cycle of generations) of the family, and various ecological conditions. It is in his analysis of how the labour-consumer balance is influenced by these various factors that Chayanov employs the concepts and techniques of marginal utility analysis, although he rejects the system of quantitative relationships (between wages, interest, rent, and profit) of conventional economic theory.

[25] Ibid., p. 6.

[26] Cf. Thorner, "Chayanov's Concept of Peasant Economy", in *The Theory of Peasant Economy,* op.cit.; and Kerblay, "Chayanov and the Theory of Peasantry as a Specific Type of Economy" in Shanin, ed. 1971, op.cit.

[27] Thorner in Chayanov 1966, p. xviii.

[28] Chayanov, op.cit., p. 5.

[29] Ibid., pp. 11f.

Conclusion

The conceptual importance of *Chayanov's* position is his insistence on peasantry as a specific type of economy even if the basic decision-making approach of marginalist economics is retained. This is closer to the formalist position in economic anthropology than to the substantivist analysis of different modes of institutional integration of the material economy. Chayanov analyses the particular economic operation of the peasant family farm as it is affected by demographical and ecological factors; he is not preoccupied with different institutionalizations of the economic sphere. His picture of peasant economy is rather static, mainly applicable to thinly populated countries where new land can readily be taken in by the farming population.[30] This picture also omits the transition towards other types of social formation.

Maximization theory might be applicable to certain problems of decision-making and choice in a peasant economy, given the institutional framework, but it obliterates the generic quality of peasantry as such. The free market model of a *contractual* society is given universal application, and economic phenomena are explained by individual behaviour–irrespective of the specific *historical* conditioning of that behaviour.[31]

The *substantivist position* of Polanyi gives a fruitful typology of principles of economic integration, but the circulation of commodities is left unrelated to modes of production. There is no general principle relating the social structure to the economy, since the economic sphere might be 'disembedded' from the non-economic system.

Marxian *structuralisme,* on the contrary, hinges historical evolution from one type of social formation to another in the economic sphere, which is given universally decisive priority over socio-political relations. This definite ontological hierarchy seems, however, rather like a decisionistic postulate.[32]

It is, on the whole, problematic to regard the *economy* as a universally well-defined field, whether it is given a status *basic* to society or it is subject to uniform principles of calculation. The modes of production and material distribution are rather points of intersection for multifarious conditions, be they of a technical, ecological, social, and

[30] Cf. Thorner, ibid.

[31] This point is stressed by Meillassoux, "From reproduction to production", op.cit., pp. 94–95.

[32] By 'decisionism' we here mean controversial positions which are assumed without any external criteria of argument or any reasonable basis of a higher order.

political kind. The character of economic 'facts' is then defined in relation to the particular mode of production in question, where the *economy* is a system of social interrelationships.[33]

We shall subsequently discuss the conceptual quality of peasantry as a historical process.

[33] For the analytical implications of this, see Lublinskaya, "The contemporary bourgeois conception of absolute monarchy", *Economic and Society,* Vol. 1, no. 1, 1972, and especially Littlejohn's introduction, ibid. Cf. also Hobsbawm's introduction to *Precapitalist Economic Formations,* op.cit., and Elster, *Økonomi og Historie,* Pax, Oslo 1972.

CHAPTER 3

Peasantry as a historical process

Everything in society might be said to be continually changing, although some types of change are generally supposed to be more 'fundamental' than others. We thus need a fixed point, by way of a theoretical construct, from which to perceive change. This 'point of view', the criteria by which change is defined, is just as controversial as the conception of any social entity.[1] Some implications will become clear through the following discussion.

The actual dynamism of peasantry

The conceptual and theoretical discussion in the previous chapters does not clarify whether peasantry should properly be considered as a rather stable and unique entity or as a mere intermediate group in a transitional stage between (pretendedly) more genuine economic formations (the 'primitive' tribal stage and 'modern' industrial farming); i.e. whether the peasant as such should be treated as a static or as a dynamic phenomenon. In our view, this is an ill-founded question. All peasantry of course exist as a historical process, but the potential dynamism inherent in all social entities might be cyclical, or it might be a continuous reaffirmation of established social patterns, and it might not lead to structural transformation of the given societal balance. Thus, given certain conditions both internally and in its external relationships, a peasant community may be rather static over a long period of time. In fact some peasant societies have retained certain basic characteristics for thousands of years. If peasantry is a transitional phenomenon, characterized by dynamic structural change and adopting a fundamentally different position in the wider society, this is, relative to a certain point of view, an empirical fact that should be accounted for in specific historical terms–it is no conceptual characteristic. To

[1] Cf. A. Pizzorno, ed. *Political Sociology*, Penguin 1971, Introduction, pp. 16–17. Pizzorno contrasts the conceptions of change of Marx and de Tocqueville to more 'pre-theoretical' conventional views. The most extensive modern discussion of such problems is Ernest Gellner, *Thought and Change*, Weidenfeld and Nicolson 1964.

clarify the implication of this in concrete terms we will briefly sketch some general traits of development, or parallel patterns, in the actual agricultural world as it is gradually more firmly determined by an urbanized, city-centred society.[2] Firstly, there seems to develop what may be called a 'cumulative pauperization' of peasantry in present-day underdeveloped countries, partly affected by a population explosion (though only under certain conditions, as *Boserup* effectively argues[3]), and partly by a growing dependence on a fluctuating market, with traditional peasant handicraft out-competed by industrial goods.[4] Increasing misery and 'hidden' rural unemployment may result when industrializing is inadequate to absorb the surplus labour population.

Secondly, the competition from large-scale, capital-intensive, and mechanized agriculture in a developed market economy destroys the small farms by making them unprofitable. This process of land-holding concentration, and closing down of farms, is highly acute in the industrialized parts of the world.[5]

Thirdly, and closely connected to the preceding point, the development of a city-dominated market economy gradually transforms the peasantry into a stratum of professional farmers, producing for reinvestment and profit through the market. This professionalization of agriculture is a consequence of specialization both of the farmer role and other roles and occupations in the countryside, as the peasant production for subsistence implies a far more diffused and unspecialized kind of work and economic adaptation. Through this process the poorest peasants may migrate from the land, while the middle peasants often manage to make a living of market production by intensive family farming, co-operative associations, and, in the more marginal cases, by hiring out their labour as worker peasants, permanent or seasonal.[6] This process had its origin in the mercantilization of agriculture in post-feudal Europe.[7] It is still not fulfilled in all parts of the industrial world, and it is blocked in many underdeveloped parts of the world, owing to a combination of commercialization and extra-

[2] Cf. Shanin's discussion in "Peasantry as a Political Factor", op.cit.

[3] Boserup's main thesis is that population growth is the main force to bring about agrarian change and technological development; see Ester Boserup, *The Conditions of Agricultural Growth,* Allen & Unwin, London 1965.

[4] S. H. Franklin argues generally that the destruction of older industries and handicrafts by the centralizing effects of the industrial revolution led to a simplification of the economy of the rural areas, called an 'agriculturalisation of the Campagnes'; cf. Franklin, "Reflections on the Peasantry", *Pacific Viewpoint,* Vol. 3, No. 1, March 1962, pp. 1–26, see especially pp. 6–7.

[5] Cf. Clout 1971 and Franklin 1971, op.cit.

[6] Cf. Galeski in Shanin, ed. 1971, op.cit.; Franklin, *European Peasantry: The Final Phase,* Methuen, London 1969; Ottar Brox, ed., *Norsk jordbruk. Utvikling eller avvikling?,* Pax, Oslo 1969.

[7] Cf. John Hicks, *A Theory of Economic History,* 1969 passim.

economic coercion on the peasantry.[8] Thus the major pattern of change which started in 18th and 19th century Europe, and which dominates the contemporary world, involves a progressive elimination of peasant social structure.[9] This observation does not exclude the existence of social mechanisms sustaining the established structure and limiting the impact of factors of change; i.e. selective emigration, social mobility, or power-groups blocking change.[10]

The ambiguous progress

The process of peasant integration into society as a whole has been seen as implying a fundamental paradox in the situation of the peasantry.[11] Technical modernization, cultural diffusion, and integrative communication systems draw the peasantry into the national society and strengthen their possibilities for political articulation at the national level. The clue to this (potentially) favoured position in an integrated nation-state is sheer political weight. But at the same time this emancipated position is threatened as the technical improvement that has transformed agriculture is also diminishing the proportion of the population engaged in agriculture, and industrial competition may render the peasant economically useless.[12] This developmental ambiguity should be kept in mind, making for a certain asymmetrical relationship in the links to the outside world, where 'the peasant view of the bad life' (Bailey 1966), its noted suspicion and traditionalism, may easily be interpreted as a highly rational defensive reaction.

The transitional dynamic

According to the specific character of peasant economy, the subsistence-oriented family-labour farm, a transition from peasant cultivation to capitalist farming is not only a gradual change in socio-economic situation, but marks a qualitative transformation from one type of agrarian economy to another. In the historical context of 19th-century Scandinavia, peasant economy is, we believe, most fruitfully considered as a dynamic phenomenon, more dependent–in its internal functioning as well as its external relations–on the development of a capitalistic environment.[13]

[8] Cf. Wolf 1955, op.cit.; Ernesto Laclau in *New Left Review*, 1971; Ernest Feder, *The Rape of the Peasantry*, New York 1971.

[9] Cf. J. M. Halpern and J. Brode, "Peasant Society: Economic Changes and Revolutionary Transformation", in *Biennial Review of Anthropology*, Stanford 1967, p. 76.

[10] Cf. Shanin, "Peasantry: delineation . . .", op.cit.

[11] Cf. Wolf 1966, op.cit.; Hicks 1969, op.cit.; Shanin 1966, op.cit.

[12] Cf. Hicks 1969, p. 120; Wolf 1966, p. 12.

[13] Cf. Kerblay in Shanin, ed. 1971, p. 159.

The generalization of market relationships seems to stimulate a split of the unified duality of the household, as the unit of production is separated from the unit of consumption. But there is not *necessarily* a positive correlation, as a way of production with unified units of production and consumption and a market type of circulation is possible: the peasant family farm producing for and dependent on the market. This situation, however, is not a stable one; it implies some disintegrative tensions in the socio-economic adaptation, the outcome and peculiarity of which should be specified in concrete structural and historical contexts. As far as social residues, institutional relics of past historical situations, may also make their mark on processes of change, these processes are not characterized by any universally predictable linearity, but by specific types of what *Lefebvre* once called 'vertical complexity'.[14]

Some general and tentative remarks about the impact of market development may be appropriate at this point.

The exchange centres typical of conditions in which major parts of the produce are consumed by the production units themselves are the local market-places. These may be necessary for obtaining complementary goods and services, like spices, salt, tobacco, or ceremonial requirements, and are usually characterized by personal relations in 'sectional' group monopolies.[15] Several factors may bring about the integration of peasant communities into wider impersonal network markets, like new tax claims, population increase, capital investment from outside, or debts to commercial middlemen through increasing economic claims.[16] The rise of market economy has far-reaching implications for the economic adaptation of the peasantry, as for instance kinship is likely to lose some of its actuality. When corporate communities are opened up to the universal money economy, and produce is transformed into merchandise, community relations tend to be tranformed into single-interest relations of individuals with goods for sale.[17] At the *micro-level* a growing inclination towards entrepreneurial calculation will be required, accompanied by further specialization of farm work, further social differentiation and division of labour in the rural community, and further individualization as family and kinship ties lose some of their economic meaning–kin dependence may be growingly replaced by wage earners. Thus characteristic aspects of market relations, like universality, anonymity, and monetary profit,

[14] Henri Lefebvre 1953, op.cit.

[15] Cf. C. Belshaw, *Traditional Exchange and Modern Markets,* Prentice-Hall 1967: Wolf 1966, op.cit., pp. 37–50.

[16] Cf. Forman and Riegelhaupt 1970.

[17] Wolf 1966, p. 48.

seem to contrast with the typical way of living of the peasant community.

At the *macro-level* the growth of the market not only implies integration in the wider society, but it implies that agriculture loses its role as the dominant mode of production, and is left behind in a secondary place. Accumulation of capital increasingly takes place in the dominant industrial and financial sectors, to which agriculture becomes subordinated, but to which it provides urbanized labourers, drained off from the population surplus of the hinterland–this being the economic side of the socio-political dilemma observed as 'ambiguous progress' above. These general impacts of market development mean new institutional premises for decision-making, a changing context of choice and articulation of needs. The process of market integration may be long and uneven, socially and geographically, and it may provoke different types of reactions, according to which interests are at stake or which alternative opportunities are available. The transitional process may follow different routes, according to how alternative societal configurations are reflected.

Conclusion

In our context, the criteria by which change is defined are the intrinsic character of peasant economy (family labour, household subsistence, and marginal market integration) and the relations of the peasantry to the wider societal order. The preceding discussion might be summarized as follows:

Peasantry should not be *defined* as a dynamic rather than a static phenomenon. Conceptually it has a generic quality which might be historically stable as well as fluctuating, and it is no mere intermediate group in transition between more 'genuine' social entities.

The transitional dynamism is, if present, a rather contingent fact which should be accounted for in historical terms. The generalization of market relationships seems to involve a change in the identity of agricultural groups, with peasants tending to assume the quality of 'farmers'. The accentuated penetration of money economy into the rural areas thereby brings about a structural change of peasantry as such. This implies both the acquisition of new type characteristics and a strategical dilemma in relation to the wider society.

Next, we will consider the response of peasantry itself: the study of peasantry as carriers of change, as 'social forces' provoking change and giving it direction and meaning. The various perspectives on peasant politics, peasantry as a political agent or victim of social coalitions, are our subsequent topic.

CHAPTER 4

Peasant Politics and Agrarian Structure

In general terms our problem was stated as the relationship between agrarian structure, processes of economic change, and political articulation (cf. the Introduction). To spell out somewhat more clearly the implications of our approach, it seems useful to relate it to previous studies focusing on the relation between social structure (as constructed by various relationships to land as a factor of production), processes of change, and political consequences. Works of reference are optional, but it seems reasonable to concentrate on Barrington Moore's *Social Origins of Dictatorship and Democracy,* which is the modern classic in this field. In a later section of this chapter we will survey the modern studies of the political mobilization of peasantry which share our own focus on peasant politics rather than Moore's emphasis on the political system itself.[1]

It should be noted that the choice of optional works is not motivated by any striking resemblance to the *substance* of our own topic, but rather by *formal* similarities, in as much as they are comparative investigations of the political autcome of economic change in various agrarian settings. The relevance to our own subject of their problem definitions and theoretical perspectives will be spelled out.

When peasant politics is related to socio-economic development, this presents, as it were, 'the other side of the coin', the complementary reverse of a conceptual specification of peasantry: the focus is shifted towards the *diversity* of peasant societies, their different historical roots, and different structural contexts. The preceding chapters have been the starting points for an analysis of peasant society as a qualitatively distinct type of socio-economic organization. The conceptual generalization aims at being the way to pin down comparative heterogeneity–which will be exposed below.

[1] In chapter 8 we will also, however, outline the systemic context of peasant action.

The lord-and-peasant nexus

Barrington Moore's effort might be concentrated in a sweeping statement, by saying that he focuses on the macro-political outcome of various ways of introducing commercial agriculture under different structural conditions. His concept of 'modernization' implies an economic and a political aspect combined: the intrusion of the market, and the process of centralized nation-building. Reflecting various lord-and-peasant constellations there are different routes to the 'modern world'. One route is what Moore calls a bourgeois or capitalist democracy, as exemplified in the cases of England, France, and the United States. Another route is the authoritarian dictatorship of fascism, as it emerged in Japan and Germany. And the third road in Moore's typology is communist regimes based on peasant revolutions, as established in China and Russia. To this selection of polities Moore adds a more undecided case, India, where he tries to formulate the alternative options available in terms of his basic analytical framework.

This framework is here our principal interest, and Moore concentrates what he endeavours to explain: 'To sum up as concisely as possible, we seek to understand the role of the landed upper classes and the peasants in the bourgeois revolutions leading to capitalist democracy, the abortive bourgeois revolutions leading to fascism, and the peasant revolutions leading to communism. The way in which the landed upper classes and the peasants reacted to the challenge of commercial agriculture were decisive factors in determining the political outcome' (Moore 1967: xvii). In Moore's full-fledged analytical model there seem to be five major variable relationships:[1]

1) The relationship of the landed upper classes to the central dynasty and its bureaucracy.
2) The response of the landed aristocracy to the requirements of production for the market.
3) The relationship of the landed aristocracy to the urban commercial bourgeoisisie.
4) The link between the dependent peasantry and the land-owners or rural overlords.
5) The structure of the peasant society; on the one hand, the class divisions in peasant society, and on the other the degree of solidarity and cohensiveness in the peasant community.

[1] The 'model' is not so formalized; this is an interpretation, and the number might well be expanded or reduced.

In these five major relationships there are four sets of actors: the central dynasty and its bureaucracy; the landed aristocracy; the urban bourgeoisie; and the peasantry. None of these sets of actors have proved to be strong enough to constitute the modernizing agent on its own, setting all the others aside. They have had to build coalitions and alliances as carriers of change or in protection against change. Of the possible alliance options–triples (three-against-one), pairs (two-against two), or all four in harmony–Moore sets out four specific coalitions for detailed investigation, as the different combinations can be styled crucial in the historical formation of the polities chosen.

Firstly, a strong coalition between the landed aristocracy and the central dynasty/bureaucracy, with a weak bourgeoisie and a highly subordinate and dependent peasantry has favoured the fascist solution, with modernization from above (Prussia, Japan).

Secondly, a modernizing coalition between the commercial bourgeoisie and the landed aristocracy, with a weak central bureaucracy/sovereign and a dispersed and transformed peasantry has favoured the bourgeois democratic path–as indicated by the English case. After the purge of the Civil War this pattern is supposed to hold even for the United States.

Thirdly, a coalition between landed upper class, bourgeoisie, *and* sovereign/bureaucracy can also build a modernizing alliance within a bourgeois-democratic framework, as in the case of France. But this was attained through a bourgeois revolution, and has resulted in an unstable combination of plebiscitarian rule and fragmented parliamentarianism.

Finally, a coalition between a new central bureaucracy (growing out of a revolutionary party) and the peasantry constituted the basic alliance which characterized communist rule, as in Russia and China. Both had a weak or non-existent bourgeoisie, and the established 'agrarian bureaucracy' and the rural overlords were the main obstacles.

If this, highly schematically, is the formal pattern, Moore treats the process of commercialization as the mechanism between social structure and types of alliances, on the one hand, and political consequences on the other. His crucial point is that the way in which the intrusion of the market is introduced, and how commercialization is responded to, determines what he calls the 'solution to the peasant question'. We may illuminate this point by four historical cases, central to Moore's argument:

In *England* commercial farming was taken up by the landed aristocracy itself, and the peasants were more or less set free to shift for themselves. In fact Moore regards the enclosures as the final elimination of the peasant question–controversial as this specific point might be.[2] And when the peasants were eliminated, there was no massive reservoir of peasants to serve the reactionary ends of the landed upper classes (as in Prussia and Japan), nor a mass basis for peasant revolution (as in Russia and China).

In *France* the landed elite took up commercial farming *through* the peasants, who *de facto* possessed the soil, and who turned a share of the produce over to the nobles, who sold it. Along with this, the power of the pre-commercial landed aristocracy was broken (in contrast to Prussia and Japan) by the 1789 Revolution. The emerging 'modernizing coalition' of central bureaucracy with both the landed and the urban interests left out weak and small-holding peasantry.

In *Eastern Europe* free peasants were reduced to serfdom, especially by the junkers of Prussia, in order to grow and export grain. Through strongly labour-repressive agriculture the peasantry was then controlled by the reactionary coalition of landed elite and central bureaucracy.

The *United States* never had a pre-commercial agrarian structure or a dependent peasantry. Moore characterizes plantation slavery as the most extreme form of repressive adaptation of capitalism, while the landed elite was broken by the Civil War and dominated by the industrial bourgeoisie.[3]

The bourgeois revolution is then a fundamental prerequisite for capitalist democracy. This bourgeois and commercial breakthrough is either supported and promoted by the landed upper classes, as in England, or, alternatively, it destroys them, as in France. This is connected to the solution to the peasant question; in Moore's words: 'Indeed the English experience tempts one to say that getting rid of agriculture as a major social activity is one prerequisite for successful democracy. The political hegemony of the landed upper class had to be broken or transformed. The peasant had to be turned into a farmer producing for the market instead of for his own consumption and that of the overlord'. (1967 : 429.)

These varieties in commercialization, bourgeois or fascist, are

[2] An interpretation of the historical material which drastically reduces the importance of the enclosures, opposed to for instance the Hammonds' classical *Village Labourer* (1912), is Chambers & Mingay, *The Agricultural Revolution 1750–1880,* Batsford, London 1966, chapter 4. Moore is closer to the Marxian interpretation, stressing how the enclosures made a substantial rural proletariat. A more balanced presentation of the opposite 'revisionist' interpretation is given by Mingay in a later pamphlet, *Enclosure and the Small Farmer in the Age of the Industrial Revolution,* Studies in Economic History, Macmillan, London 1968.

[3] It seems quite misleading, however, to view the society and economy of the Slave South as an adaptation of *capitalism.* Slavery is a social system not primarily based on reinvestment and maximization of profit, as effectively argued by Eugene Genovese; see especially *The Political Economy of Slavery,* Random House, New York 1965.

contrasted by the experience of Russia and China, which stresses the significance of the *failure* to provide an appropriate form of commercial agriculture, promoting peasant revolutions because the landed upper classes did not manage to turn to production for the market in a way that enabled commercial influences to permeate rural life (cf. Moore, p. 459). The social structure did not permit the adequate centre-forming and modernizing coalition based on existing dominant groups.

The outline above may be said to present the major structure in Barrington Moore's analysis, his main factors and categories of explanation, in a highly simplified form. This structure of argument is quite smooth and elegant in the preface and the extensive concluding chapters. On no account does it do justice to Moore's detailed discussion of the individual historical cases. The complications in the historical analysis might also, however, have some serious consequences for the general argument itself. In this context we will not question historical exposition or points of fact, even if specialist historians, in one part or another of the vast field which 'Social Origins' covers, may quarrel over accounts or interpretations.[4] To clarify its *inherent problems,* we will rather question the book on its own premises.

Mainly, Moore follows his own program in the analysis. But in a wide range of important questions he also seems to rely on *ad hoc* hypotheses and explanations. As factors behind the different possibilities for introducing commercial agriculture, for instance, he merely gives a general summary, without systematic introduction of the elements into the analysis: the existence or non-existence of commercial markets in a near-by town; variations of soil and climate; what is called 'political considerations', such as, inter alia, if the landlords choose to make use of the coercive apparatus of the state to collect rents (p. 422). Sometimes he uses ideological explanations of such 'political considerations', while his main preference is to specify them as functional adaptations to a changing socio-economic environment. Sometimes he even finds it necessary to rely on factors external to the society under consideration, as when the explanation of the actual mobilization of the Chinese peasantry is firmly based on Chalmers *Johnson's* stress on the importance of the Japanese occupation.

[4] Some examples are given by Stanley Rothman in a rather hostile review article, "Barrington Moore and the Dialectics of Revolution", *American Political Science Review,* Vol. LXIV, No. 1, March 1970, pp. 61–83. Rothman's major accusation is that Moore stretches the historical evidence by a 'predisposed' political standpoint of 'neo-Marxist' flavour.

Generally, his relations to non-economic factors, and their theoretical status vis-à-vis economic ones, seem to be a bit problematic and undecided: the irresolution may be connected to combining an attempt at being a genuine Marxist with a striking fear of the accusation of being one.[5] The exposition may then somehow be a mirror of this slight neurosis. The main solutions in national polity, however, are given a clear economic background, while cultural, linguistic, and religious factors only marginally enter into the discussion.[6] He discusses the importance of religious tradition in the formation of elite alliances and peasant responses in India, Japan, and China, but concludes that 'the style of center-formation' (Rokkan 1970), or obstacles to centre-formation, are more important than rituals and belief systems: the social segmentation through the caste system in India, the vulnerability to peasant mobilization due to concentration of power in a central 'agrarian bureaucracy' in China, the fragmentation of peasant solidarity through a patrimonial system of control in Japan. Control and manipulation of symbol systems is only given a marginal and *ad hoc* status.

To sum up, this periodic introduction of *ad hoc* explanatory factors outside the main model is to some extent legitimate and necessary. But when there is an apparently unsettled variation of factors from case to case, it is somewhat difficult to get a firm grip on the essence of the *comparative* insight.

The slight confusion might be due to some indistinctness in theoretical equipment: one explicit basis for change is social structure, the dependence relations and coalitions as responses to and origins of change. Another just as explicit basis for 'Social Origins' is the focus on carriers of change, how they try to optimize their benefits, in exploitation or defence, what they expect of the change they carry, and what they get out of it. Moore spells out the *relations* between lord and peasant, between landed nobility and bourgeoisie, between rural classes and peasant community groups, and he uses the various configurations in his explanatory scheme. But as a central force he also focuses on the intentions and conscious manipulations of the dominant classes, their strategies for benefits and effective exploitation. He does not, how-

[5] Ronald Dore pictures the outcome of this internal struggle (inside 'the Bishop of Woolwich of the Marxist school') as follows: "he has both feet firmly planted in the mainstream of the Marxist tradition. To be sure there are times when the holy water hardly seems to be up to his ankles", in "Making sense of history", *Archives eurpeéennes de sociologie,* X, 1969, p. 302.

[6] Cf. the extensive discussion by Stein Rokkan, "Methods and Models in the Comparative Study of Nation-Building", in Rokkan, *Citizens, Elections, Parties,* Universitetsforlaget, Oslo 1970, especially pp. 54–60.

ever, spell out the relationship between a perspective oriented towards 'structure' and one oriented towards intentional actors.[7]

But even if 'Social Origins' is not totally successful (how could it be?) its major approach, what it *tries* to do, and partly succeeds in doing, is a path-breaking challenge in comparative social research.[8]

To what extent is this *problématique* relevant to Scandinavian agricultural change?

Spontaneously the relevance seems quite obvious: new cases offer the possibility of extended insight by a wider empirical framework. It implies investigation in a new historical setting of the role of agrarian structure, patterns of dominance, and commercialization, as generating political mechanisms.

But this prospect of pay-off is of course far too sweeping, and we face at least three important complications.

Firstly, Moore explicitly concentrates on the leading and powerful nations, and leaves the smaller and less influential polities out of consideration. He regards the smaller nations as being too dependent, since their political configurations are often generated from outside, to be comparable to the leading ones. In fact he questions the very possibility of a general model for analysis of political development, 'likely to be so broad as to be abstractly platitudinous' (p. xiii). *Contra* this view, however, it should be stressed that it is no argument against an attempt to investigate the *influence* of mercantilization and agrarian structure. If considerable external pressure is a disqualification, Moore's stress on the impact of Japanese occupation in the Chinese case would be just as dubious, and so would the implications of British colonialism in Indian history. So, the criteria for 'adequate' independence seem rather flexible anyway. Further, it should be stressed that Moore does *not* give any conclusive comparative generalizations as regards historical substance; he uses a (relatively) common framework of analysis for different national stories.

Secondly, it is doubtful what happens to Moore's scheme when central sets of actors are obviously missing. The structural model is not directly applicable to, say, Africa, so far as an urban 'bourgeoisie', as well as a State, were lacking in a crucial phase of development, nor does it seem applicable to Scandinavia (especially Norway) in its (rela-

7 This traditional distinction in sociological methodology is given a new dramatic emphasis in French *structuralisme,* especially in the labelling of intentional explanations–the use of a conscious *subject* as the moving force in history–as a refutable *historisisme;* cf. Althusser et al., *Lire le Capital,* op.cit., and Nicos Poulantzas, *Pouvoir Politique et Classes Sociales,* Paris 1968.

8 There has, however, been some doubt as the extent to which *Social Origins* is excellent historiography with somewhat troublesome sociological 'generalizations', or whether it is the other way round: exemplary sociology but rather dubious history. Dore, op.cit., holds the latter view, with some modifications.

tive) lack of a landed nobility. What should be noted is that the constellation is different from the polities analysed in 'Social Origins'.

Thirdly, our main question differs somewhat from that of Moore (and this is an optional point), as our focus on 'political outcome' stresses peasant politics rather than the political system itself. This could be just a minor shift in stress, as peasant political articulation presumably is closely connected to the general political framework (which is outlined in chapter 8), but it may nevertheless have important consequences. To illustrate, Norway and France would not be grouped together, as 'capitalist democracies', but contrasted: Norwegian peasants developed their own national political strategy, their own coordinated class policy, and finally a distinct agrarian party, while the French peasantry seems to have limited their distinctive articulation to scattered outbursts of 'spontaneous' protests. In a tentative explanation of the contrast, it seems natural to discuss the possible impact of symbolic articulation and cultural dominance; shortly: the Catholic Church. In Norway the peasant movement, mostly rather slow and gradual, was simultaneously a national and cultural awakening–against Danish influence and Danish civil servants.The economic class policy was closely knit together with a cultural struggle at the symbolic level–religion from the very beginning and language later on. Such connections were more unlikely within a centralized and unitarian Catholicism, which at the same time seems rather functional for the dominant 'modernizing' coalition.

Thus the specific impact and consequences of *commercialization* may be easier to localize within a narrower culture-specific region, which the Nordic area is, avoiding perhaps some of the ontological assumptions implicit in the global scope of 'Social Origins'.[9]

Next, we shall give a selected analysis of contemporary approaches focusing directly on the relationship between agrarian structure, the strain of economic change, and the political role of peasantry itself.

The theories of peasant mobilization

Two main theoretical traditions, employing a markedly different conceptual apparatus, have dominated the sociology of political movements, mobilization of peasantry inclusive. Both of them are highly heterogeneous, but we shall draw a line of demarcation which explicitly intends to illuminate the study of peasantry. On the one hand some theoretical schools are preoccupied with the *integration of the social system* and the *function* of institutional arrangements. On the other hand, historical and comparative approaches, Weberian and Marxist alike, focus on the struggle for the *defence of interests,* the con-

[9] This view is elaborated by Rokkan, 1970, op.cit.

stellation of social forces which defines the context of this struggle, and the political dominance which influences the outcome of it. We shall specify the analytical implications of each camp a bit further.

Social integrationism. The main type of conceptual approach to the integrative function of social systems might be labelled 'mass society theory', closely allied to Parsonian functionalism. *Kornhauser* (1959) argues that the lack of strong secondary groups, associations or organizations, mediating between elites and non-elites, stimulates 'mass behaviour' susceptible to manipulation, control, and expressive mobilization from above. He supports de Tocqueville's argument that social pluralism works best for a stable democracy. The social anonymity of mass society is supposed to dissolve social coherence and normative integration, leading to isolation and 'anomie'. Crucial concepts in mass society approach are 'anomie' (normlessness), 'alienation' ('meaninglessness' in a psychologistic interpretation), and 'disintegration' (of social cohesion). *Durkheim* is one of the founding fathers. His notion of the anomic centres on the collapse of recognized codes and communities, when the industrial society breaks up traditional statuses, thus implying total fluidity of aspirations.[10] While anomie is the abnormal state, *solidarity* is the normal one. The Durkheimian distinction between types of solidarity, the 'mechanic' solidarity of a simple preindustrial community (based on functional equality) and the 'organic' solidarity of complex industrial society (based on complementarity in occupational functions), is elaborated by contemporary students of industrialization and rural political behaviour. Erik *Allardt* employs Durkheim's conceptual framework (degree of division of labour, type and degree of solidarity, and degree of pressure towards uniformity) in his analysis of political radicalism in Finland.[11] The strong element of 'backwoods communism' in the peripheral rural communes in eastern and northern Finland is explained by mass society theory: loosely integrated communities with both low solidarity and low division of labour, socially uprooted individuals without ties to secondary groups, thereby stimulating radical protest and expressive political participation.[12] In Allardt's scheme the Durkheimian approach is combined with reference group theory, since types of social reference

[10] Cf. Emile Durkheim, *Suicide,* Routledge and Kegan Paul, Longon 1952.

[11] Durkheim, *Division of Labour in Society,* Free Press, Illinois 1947. Allardt, "Types of protests and alienation", in Allardt & Rokkan, eds., *Mass Politics,* Free Press 1970, pp. 45–57. See also Allardt, *Samhällsstruktur och sociala spänningar,* Almquist & Wiksell, Stockholm 1965. In this framework mechanical solidarity is characterized by low division of labour and high pressure towards uniformity.

[12] The crucial distinctions here are 'expressive' vs. 'instrumental' action, and 'backwoods' vs. 'industrial' radicalism; cf. Allardt 1970, op.cit.

groups are supposed to coincide with different types of solidarity.[13] 'Backwoods communism' is seen to imply *diffuse relative deprivation,* since relevant reference groups are lacking in that type of anomic society, while lower class deprivation might be *institutionalized,* rather than diffuse, in more integrated organic societies. The notion of relative deprivation, crucial in reference group theory, is also the cornerstone of the general explanation of human rebelliousness presented by Ted R. *Gurr.*[15] This variant of *integrationistic* theory has probably the strongest affinity towards social psychology, and contains important traces of psychologistic reductionism, where social relations are based on the interpersonal experience of individual actors rather than the structural character of society as a whole.

On the other hand, Parsonian theory of social action is elaborated by Neil *Smelser,* to locate the functions of systems and to trace the development of integrative institutions.[16] By the evolution of increasing differentiation, multi-functional role structures (ideal-typical of 'primitive' pre-industrial societies) are replaced by highly specialized structures.[17] Uneven advances of differentiation and integration, the obsolete character of social processes, are reflected in social disturbance like 'value-oriented' or 'norm-oriented' movements.[18] Collective action is a normative mobilization in the name of a 'generalized belief', where the defining characteristics of collective behaviour are, in Smelser's Parsonian conceptualization, social rather than psychological.[19]

In spite of the heterogeneous variety of mass society theory, from psychologistic reductionism to social systems analysis, certain common themes might be identified in the approach to collective behaviour and political mobilization. There is a general emphasis on the expressive aspects of mass behaviour, where political protest is seen either as a mere *reflex of deprivation* or as an *anomic symptom* characteristic of

[13] Societies of mechanical solidarity are, for instance, supposed to be based on similarity and strong normative attachment, lacking alternative comparison groups from out side. Thus specific class-consciousness is hindered by diffuse social identification. Cf. Allardt 1970, op.cit.

[14] Ibid.

[15] Gurr, *Why Men Rebel,* Princeton University Press 1970.

[16] Cf. Neil Smelser, "Mechanisms of Change and Adjustments to Change", in B. Hoselitz and W. Moore, eds., *Industrialization and Society,* UNESCO, Paris 1963, pp. 49–74.

[17] This perspective is also elaborated by, inter alia, David Apter, *The Politics of Modernization,* Chicago 1965, and S. N. Eisenstads, *Modernization–Protest and Change,* Prentice-Hall, New Jersey 1966. See also Ioan Davies, *Social Mobility and Political Change,* Macmillan, London 1970, chapter 2.

[18] Cf. Smelser, *Theory of Collective Behaviour,* Routledge and Kegan Paul, London 1962.

[19] Ibid., p. 313; see also, for a comment, Paul Wilkinson, *Social Movement,* Macmillan London 1971, pp. 25 ff.

culturally and socially uproooted individuals with weak ties to integrative secondary groups. The *neo-evolutionist* analysis of structural differentiation stresses the universal strain implicit in the transition from pre-industrial society to the 'modern' industrial one, and the collective mass response to this inevitable strain.

The determinism of these types of theory demands critical examination: all of them imply an extremely general approach to human behaviour irrespective of that historical specificity which structures the wider context of behaviour and the institutional framework of choice, and irrespective of those different socio-economic conditions which determine interest constellations and alliance options. So far as the historical reality is accounted for, this is made within an evolutionist scheme of general stages of differentiation. Moreover, the systems integration approach is not geared to the study of *peasantry,* and gives no clues to account for specific changes in the quality of peasantry as such. We will not rule out the possibility that mass society theory might be helpful to explain certain collective aspects of peasant politics–by no means; but other theoretical traditions have employed fruitful explanatory categories closer to the structural peculiarity of peasant societies themselves.

Interest mobilization. Theoretical perspectives stressing the interest articulation of peasantry, between subgroups of peasant society or between peasantry as a class and other social groups, often focus somewhere on the axis between *dependence* and *autonomy* in a structural relationship.[20] It is hardly correct to say that an emphasis on peasant dependence rather than on peasant autonomy represents different theoretical traditions. They are explaining different types of behaviour, and they focus on different types of peasantry. Thus they illuminate the preconditions of peasant politics from different angles.

While *Redfield* once observed that 'peasantry have been a conservative factor in social change, a brake on revolution, a check on that disintegration of local society which often comes with rapid technological change',[21] *Fanon* concluded that 'dans les pays coloniaux seule la paysannerie est revolutionnaire', that the peasant, having nothing but his chains to lose, being declassed and exploited, 'decouvre le plus vite que la violence, seule, paye'. (Fanon 1961). Between these two diagnoses, one stressing the cultural dependence of peasantry, and the other its potential of naked autonomy in a cosmic act of liberation, grows a steadily increasing literature on peasant movements and revo-

[20] A general interpretation of peasant movements in terms of 'structural autonomy' is presented by Archetti, Fossum and Reinton "Agrarian Structure and Peasant Autonomy", *Journal of Peace Research,* No. 3, 1970, pp. 185–195.

[21] Redfield 1956, op.cit., p. 77.

lutionary mobilization of peasantry, from rather specific case studies to the more grandiose high-level generalizations.[22] Substantial parts of the general debate might be a rather dubious undertaking, but a survey of the main contributions will be relevant to the present analysis.

The perspective stressing peasant dependence, the generally restricted scope of manoeuvre, is applied by authors who analyse the *resistance to change* of peasant societies as a result of cultural and political subordinacy to local lords and external authorities. Robert Redfield focuses on the cultural dependence of peasantry, based on the degree to which peasants share the values of superior groups. The dominance by superiors is based on legitimate authority rather than coercion.[23] This perspective is also elaborated by Charles *Erasmus* in his notion of the '*encogido* syndrome', the timid, withdrawn, and suspicious personality conditioned by low status and deprivation.[24] This syndrome, supposed to achieve its most elaborate form in 'closed' peasant communities, involves psychological traits of insecurity and lack of self-confidence, and is seen as generally produced by similar social conditions irrespective of cultural setting.[25] The *encogido* syndrome is thus a dependence effect resulting from social conditioning, an underdog syndrome conditioned by low status in the dominance/subordinacy hierarchy and by economic and educational deprivation.[26] As for the peasant 'resistance to change' versus its potential of rebellious action, Erasmus favours the de Tocqueville model of revolution, implying that peasants are more likely to be rebellious when conditions become less oppressive (or suppressive) than when they are increasingly suppressive as in the Marxian model.[27] 'The effects of oppressive dominance/subordinance hierarchies can lead to feelings of powerlessness and to withdrawal *(encogido)* behaviour in the subordinate stratum'.[28]

The structural dependence of peasantry is also stressed by Gerrit *Huizer,* but seen as result of a 'culture of repression', involving violence and coercion by dominant landlords and not based on shared values.[29]

[22] A substantial part of such studies seems to be confined to Latin America, see, inter alia, Rodolfo Stavenhagen, ed., *Agrarian Problems and Peasant Movements in Latin America,* Anchor Books 1970; and Landsberger, ed., op.cit.

[23] See Redfield, *Peasant Society and Culture,* University of Chicago 1956.

[24] Erasmus, "Community Development and the Encogido Syndrome", *Human Organization,* Vol. 27, No. 1, Spring 1968, pp. 65–74.

[25] Ibid., p. 71.

[26] See Erasmus' comment to a critique by Gerrit Huizer, *Human Organization,* Vol. 29, No. 4, Winter 1970, p. 316.

[27] Ibid., p. 318.

[28] Ibid., p. 319.

[29] Huizer, " 'Resistance to Change' and Radical Peasant Mobilization: Foster and Erasmus Reconsidered", *Human Organization,* Vol. 29, No. 4, Winter 1970, pp. 303–313.

The social and economic traditionalism of peasantry is interpreted as a strategic defence reaction, as an inverse expression of the 'innate rebelliousness' and revolutionary potential of peasantry. Huizer argues that a change of power relations may take place through peasant action or otherwise, and he considers land invasion to be a typical and non-violent strategy.[30] While Redfield and Erasmus thus presume the impossibility of radical mobilization of a culturally dependent peasantry, Huizer gives some strategic clues to resisting the repressive conditioning of apathy. But this potential development of political consciousness is not systematically related to structural conditions, or to economic differentiation within the peasant world.

The internal differentiation is the main focus of perspectives stressing the *autonomy* of specific types of revolutionary peasantry. This emphasis on the revolutionary potential of peasantry does not then pretend to have general applicability, but is restricted to the structural characteristics of particular peasant groups. A short discussion of two of the more outstanding contributions within this tradition can give us some clues of analytical interest.

Eric *Wolf*'s *Peasant Wars of the 20th Century* focuses on the most dramatic expression of peasant political articulation, their participation in nation-wide revolutionary upheavals. His selection of cases includes the Russian revolution of 1917, the Mexican revolution from 1910, the Maoist revolution in China, the Castroist takeover on Cuba, and the national liberation war in Algeria from 1954, and in Vietnam since the second world war. Wolf states that his main contribution to the understanding of the topic, as a social anthropologist, will be the differentiation between specific types of peasants involved; then the micro-sociological perspective, that is: the meaning of community structure, of regional variations, of local ecologies; and finally the focus on the various mediation groups between the peasants and the wider society, the landlord, the merchant, the priest, the politician. Now, it could be argued that there seems to be a slight gap between intentions and accomplishment, between preface and conclusion, on the one hand, and the description and analysis of empirical cases on the other.[31] The narrative of prelude and revolutionary events is more conventional, and not always steered by the explicit purpose: the analysis of land relationships and peasant mobilization is, as it were, less extensive. However, the special premises for the general theoretical formulations in the conclusion are quite comprehensive in that chapter itself, which then is able to stand on its own, present-

[30] Cf. Huizer, "Land Invasion as a Non-Violent Strategy of Peasant Rebellion", *Journal of Peace Research*, No. 2, 1972, pp. 121–132.

[31] Norman Stone makes a point out of this in a rather hostile review article in *The Spectator*, 9 October 1971.

ing the core of the argument.[32] Wolf's theory of the mechanisms at work in revolutionary participation of peasantry can be condensed as follows. Generally, he sees a crucial strain connected with the specific conditions in the change-over from peasant to farmer, or the intrusion of capitalist production seen as a change of the institutional context within which decision-making takes place. In his own words: 'Perhaps it is precisely when the peasant can no longer rely on his accustomed institutional context to reduce his risks, but when alternative institutions are either too chaotic or too restrictive to guarantee a viable commitment to new ways, that the psychological economic, social, and political tensions all mount toward peasant rebellion and involvment in revolution' (p. xv). This seems quite reasonable, but the significance is not quite clear, except that peasants rebel when they find it imperative to do so. In the concluding chapter, however, the institutional presupposition is specified. The main thesis is that some degree of tactical mobility or power, some degree of structural autonomy in relation to lord and market and management of plot, is a necessary prerequisite for rebellious action. This tactical mobility is achieved either through land ownership, or through location in a peripheral area. This relatively autonomous peasantry is also vulnerable to commercialization, when land and produce are made into market commodities, and they rebel against the threat to the precarious equilibrium which their tactical power makes it possible to achieve.[33] 'Tactical mobility' is thus a general precondition of rebellion, and rebellion is a reaction in defence of that mobility. But the degree of autonomy needed, and how dangerous the threat ought to be, is not fixed in universally abstract terms, and it might be rather misconceived to ask for it. The general thesis that peasants who own land or are ecologically isolated are more likely to rebel than more dependent ones, is a minimum-factor whose prognostic value seems supported by empirical evidence.[34]

Hamza *Alavi,* in a study called "Peasants and Revolution" (1965), shares (or rather precedes) Wolf's general opinion, stating that the

[32] This is also done in a previous article by Wolf, "On Peasant Rebellions", *International Social Science Journal,* Vol. XXI, No. 2, 1969, pp. 286–293.

[33] In his excellent work on *Peasants,* Prentice-Hall 1966, Wolf states that: "The perennial problem of the peasantry thus consists in balancing the demands of the external world against the peasants' need to provision their households", p. 15.

[34] A clear-cut delineation between *necessary* conditons and *sufficient* ones would be a long step further, but an extremely difficult one.

'middle peasant', in his relative independence of rural lords, is the most militant element among the peasantry. His subject is the successful peasant revolutions in Russia (was it a peasant revolution?) and China, and the smaller abortive ones in India.[35] Alavi seems to specify the structural factors in a way that supplements Wolf's thesis. His main explanation of the passive peasantry, who never rebel against their overlords whatever their material misery might be, is the vertical segmentation of the peasant community. These 'factions' of social relationships, as vertically integrated segments of the rural society, are a structural hindrance against horizontal solidarity and coordinated uproar and insurrection developing. The segmentation leads to what Alavi calls a 'paternalistic mystification', an identification of the poor and dependent peasant with his master. In his own words: 'The allegiance of the poor peasants and the farm labourers to their masters is not merely due to subjective factors such as their 'backward mentality' etc. It is based on the objective fact of their dependence on their masters for their continued livelihood'.[36] A necessary precondition for insurrection is then to break the factional structure; this is done either from below, by the peasantry, or from above, by the government.

The two main groups of factions, in Alavi's presentation, are the masters and their dependents, on the one hand, and the independent small-holders on the other. He concludes it empirically supported that the independent small-holder, the middle peasant, has been the militant element, the initial actor in upheavals.[37] As this may lead to breaks in the factional structure in the country-side, the poor and dependent peasants may be mobilized, and their participation then becomes a prerequisite for radical transformation of the social structure. As for the Chinese revolution, Alavi finds this sequence in accordance with Maoist practice, even if it is contrary to Maoist theory. The main interest to us, then, is the general pointing out of institutionalized segmentation, and also the points (preliminary as they are) about the relative independence needed to break the vertical network down.

Both Alavi and Wolf thus focus on the internal differentiation of peasant society, rather than on peasantry as a homogeneous and unique entity, on 'classes of peasants' rather than on 'peasantry as a class'.

[35] Alavi discusses, as does Barrington Moore, the revolutionary 'Telengana' peasant movement of Hyderabad in 1940. But the description of the movement, and the narrative of events, varies a great deal between the two; cf. Moore 1966, pp. 380–382 and Alavi 1965, pp. 268–270.

[36] Alavi, op.cit., in *The Socialist Register* 1965, p. 274.

[37] The initial role of peasants with some scope of manoeuvre is also stressed by Hobsbawm, "Peasants and Rural Migrants in Politics", in Claudio Veliz (ed.), *The Politics of Conformity in Latin America,* Oxford University Press 1967.

The dimensions of dependence and autonomy

The dependence and autonomy defining the material and social existence of peasantry is no universal mode of being. The degree of autonomy, as a prerequisite for tactical mobility, is more or less prevalent in different aspects of the social situation of peasantry.[38] We might, tentatively, single out five dimensions as highly relevant in this respect:

1) Political status of peasantry
2) Security of land occupancy.
3) Degree of managerial responsibility.
4) Relation to vertical market.
5) Family and community orientation.

'Dependence' and 'autonomy' can be employed to characterize each of these dimensions, and any combination of a high degree of autonomy along one dimension and low along others is possible. Analytically, each dimension should then be regarded as independent of the others, although some combinations might be more empirically widespread.

By *political status* of peasantry we mean the extent to which the peasants constitute a mere subject group in society, or to which they play an independent political role: if they have the vote and if they have access to the relevant political institutions; if they are effectively involved in the public decision-making process; if they are favoured by the operations of the political system, and if they benefit from the social and educational institutions of the state.[39]

The *security of land occupancy* concerns the predictable stability of tenancy conditions, leasehold, and ownership. This security may be precarious for social reasons (tenant peasantry subject to dominant landlord power without institutional stability) or for ecological reasons (frontier squatters dependent upon low attractivity and low central control of marginal land).

The *degree of managerial responsibility* describes the peasant control of the distribution of resources, the independence in time-phasing and management of their own plot. Any peasant, working close to the soil, decides *when* things should be done–when to plough, plant, harvest, milk, etc.[40] These decisions might be guided by custom and tradition, but the peasant himself has ultimate responsibility. But a more crucial measure of dependence and autonomy is the responsibility for decisions about *what* should be produced–the mixture between livestock hus-

[38] Cf. Landsberger 1969, op.cit.; Archetti, Fossum and Reinton 1970, op.cit.; Hobsbawm 1967, op.cit.

[39] See Landsberger 1969, op.cit., p. 7.

[40] This is discussed by John Hicks 1969, op.cit., pp. 115 ff.

bandry and corn production, between arable and pasture, the relation between subsistence and cash crop production, the phasing and content of crop rotation, the degree of mechanization, the use of fertilizers. The more dependent the peasant is, the less is his responsibility for decisions like these.

By *relation to the vertical market* we mean the degree of access to commercial markets outside the local marketplace and independent of landlord control. Lack of personal dependence will increase the impersonal market dependence, where market production may be stimulated by rent, taxes, or debt charges. Exploitation of surplus by landowners or by dominant commercial middlemen can make the net price received by the peasant less favourable than that received by others.[41]

The *family and community orientation* influences the extent to which peasants individualize their managerial and political actions. It influences the extent to which peasantry acts as a cohesive entity in itself, suspicious or hostile to outsiders, and the extent to which groups of peasants may split off to form their own organizations in opposition to power groups within peasant society itself. The closely knit family household unit may also optimize the subsistence exploitation of available resources, irrespective of criteria for profitable business management, although the same socio-economic organization may restrict the entrepreneurial options available. Social mobility is probably not stimulated by strong familial and communitarian networks, but is rather counteracted by levelling mechanisms.

The interrelationship between these dimensions or aspects of autonomy and dependence should be left for empirical discoverage. So should their connections to characteristics of the wider societal order, and their consequences for the articulation of peasant politics and mobilization. They are structural clues to comparative investigation rather than an elaborate model of peasant behaviour. On a highly general level it seems likely that peasant mobilization occurs in defence of structural autonomy, to keep it, to regain it when it is lost, and to gain it in order to improve the position vis-à-vis townsmen and rural lords.[42] Peasant mobilization may thus be both offensive and defensive as to the various dimensions on the scale of independence. But this implies such general assumptions about human behaviour that it should not be advocated without clearer specification of the particularity of

[41] An extensive discussion of such relationships is given by S. Forman & J. F. Riegelhaupt, "Market Place and Marketing Systems: Toward a Theory of Peasant Economic Integration", *Comparative Studies in Society and History,* Vol. 12, No. 2, (April 1970), pp. 188–212.

[42] This is the argument advocated by Archetti et al. 1980, op.cit., p. 192.

peasantry and the concrete effects of independence on peasant life. As Wolf, Alavi, and Hobsbawn convincingly put forward, autonomy, or tactical mobility, is a prerequisite for rebellious peasantry. But peasant action might be more peaceful and nuanced than that, and we should consider the general prospects of alternative political and pseudo-political articulation.

The scope of peasant politics

We have previously discussed the variety of theoretical traditions to account for revolutionary peasantry, rebellion against the established order, and against the rules of the game. This is only one type of political articulation, although the most dramatic one. There is a much broader range of alternative options, of peasant responses to the challenge of change, which might be analysed within corresponding theoretical trends.

We shall present a rough typology of peasant politics. This constitutes, though preliminarily, and outside revolutionary guerrilla warfare, the main options that have to be spelled out and explained wthin any framework.[43]

Peasant politics might be characterized by its position in relation to the following set of binary opposites:[44]

Independent	– Guided
Spontaneous ('expressive')	– Strategical ('instrumental')
Exclusive	– Participant
Non-political (or 'pre-political')	– Politicalized
Individual	– Collective

Independent political action implies a situation where the peasants, through coordinated class action, build up their own regional or nationwide organizations, with their own leadership, their own ideology and purposes. In *guided* political action peasantry is 'mobilized' by an external elite for purposes outside the scope of the peasant horizon.[45]

Spontaneous response means amorphous political action, either as sabotage and passive resistance, or as 'jacquerie'–expressive out-

[43] Except, of course, theoretical positions that deny the fruitfulness of any *typification* of political reactions.

[44] Some elements of this typology are given by Landsberger 1969, op.cit., Shanin's "Peasantry as a Political Factor", op.cit., and A. Q. Obregon, "Contemporary Peasant Movements", in Lipset & Solari, eds., *Elites in Latin America,* Oxford University Press 1967.

[45] Independent and coordinated class action is, with some modifications, *the* peasant politics in Scandinavia, in Norway from about 1830 onwards. Guided action seems more negligible on the Scandinavian scene, but covers, inter alia, the Maoist 'long march', the role of the Russian cossacks, and French bonapartism.

bursts rather than deliberate strategic coordination.[46] *Strategical* action is rather an instrumental calculation of adequate means adapted to the actual or anticipated reaction of antagonist forces.

By *exclusive* peasant politics we mean action restricted to the mobilization among peasantry themselves, while *participant* behaviour implies peasant participation in wider non-revolutionary movements, for instance those of a nationalist kind (inter alia, the fusion of peasantry and liberal bourgeoisie), and often with a populist flavour.[47]

Non-political peasant movements constitute alternative channels of articulation not directly acting upon social structure or economic conditions. Certain religious movements, millenarian and messianic, are such pre-political 'archaic forms' (Hobsbawn 1959). So are some racial movements and, above all, social banditry.[48] *Politicalized* articulation might be of different forms, from reformist legalism/constitutionalism via political banditry, epidemic or endemic, to revolutionary agrarianism.[49] The ideological myth might well be nostalgic, longing for the restoration of some golden past.

Finally, the peasant response to change might be either *individual* or *collective*. The trend towards agrarian individualism finds its historical expression in enclosures of common land and the dominance of individual sovereignty over collective obligations.[50] Geographical migration may also be counted as an individualization of peasant reactions to change.

As for the analytical status of these pairs of potential attributes, three points should be noted:

Firstly, the qualities of peasant action as specified by any of the binary opposites are not mutually exclusive. An actual peasant movement will be identified by a cross-combination of attributes from all the pairs in question, even if some combinations might be empirically more prevalent than others.

[46] European examples may be the pre-industrial food riots in France and 'Captain Swing' and 'Rebecca's Daughters' in Britain (cf. Hobsbawm & Rudé, *Captain Swing,* Lawrence & Wishart 1969, and Rudé, *The Crowd in History,* Wiley & Sons 1964). In the North this type belongs principally to the phase prior to the 'agricultural revolution'.

[47] Cf. Peter Worsley's 'rural idiom' in *The Third World,* Weidenfeld & Nicolson 1967, and Ionescu & Gellner, eds., *Populism,* Weidenfeld & Nicolson 1969.

[48] In Hobsbawm's words: "*Social banditry,* a universal and virtually unchanging phenomenon, is little more than endemic peasant protest against oppression and poverty: a cry for vengeance on the rich and the oppressors, a vague dream of some curb upon them, a righting of individual wrongs. Its ambitions are modest: a traditional world in which men are justly dealt with, not a new and perfect world." *Primitive Rebels,* Norton, New York 1959, p. 5.

[49] Cf. Obregon 1967, op.cit.

[50] A basic work here is Marc Bloch, *Les Caractères originaux de l'histoire rurale francaise,* 2 vols., 2 ed., Paris 1955–56.

Secondly, the set of binary opposites represents pure analytical types. There is in fact a gradual and continuous scale between the opposite attributes, making the calculation of a fixed number of types empirically meaningless.

Thirdly, there is here no presupposition of a universal developmental pattern, but there is rather a non-linear and irregular fluctuation between different trends. There is hardly any universal dynamism inherent in all types of peasant politics as such and irrespective of what happens in the national society as a whole.

The qualitative character of peasant politics depends, we assume, on the agrarian structure in question, the type of economy and society which constitutes the framework of social coalitions and antagonisms; it depends on the actual processes of change, on what happens to the social groups, who is affected by transitional forces, and how they are affected; and it depends on the cultural and ecological setting which defines the alternatives and the cognitive horizon of choice. This constitutes the historical situation within which options are defined (opened up or closed), and within which action takes place. The explanatory theoretical construct should be regarded as an interrelated whole rather than a mere multi-factorial list, but the analytical integration should be demonstrated through the empirical inquiry itself. Models of the political articulation of peasantry on too generel levels are easily diagnosed and refuted as platitudinous abstractions.[51]

[51] This is what Barrington Moore (1967) seeks to avoid – in our view successfully.

PART II:

Agrarian Society and its tranformations

This part is an empirical investigation of agrarian relationships in Scandinavia preceding, and during, the rather dramatic transformation in the 19th century. In chapter 5 we explore the basic figures of population growth, and discuss land tenure systems and disposal of land according to the conceptual distinctions made in the introduction and the preceding part of the book. In chapter 6 we focus on the individualization of agriculture by transformed land tenure and increasing peasant proprietorship, while the development of a substantial class of landless persons is viewed as an aspect of this change. In chapter 7 we then trace the comparative features of technical innovation and mercantilization of agriculture–the commercial and financial penetration of landed relationships. This development is finally related to those aspects of peasant autonomy which were outlined in the theoretical discussion in Part I above, and it provides the frame of reference for the analysis of peasant politics in the subsequent Part III.

CHAPTER 5

Agrarian structure of the old peasant society

Political and geographical setting

The three kingdoms of Norway, Sweden and Denmark had been united in 1397, on the diplomatic initiative of the Norwegian queen Margareta, daughter of king Valdemar of Denmark.[1] The decline of Norway during the late Middle Ages made the country unable to maintain any equality of status within the union, and from the early 16th century Norway came under the overlordship of the Danish rulers. The Norwegian state–under a continuous personal union–lost its independence and national sovereignty, and so did the Norwegian church (with the Reformation of 1536).[2] Sweden, on the other hand, achieved its national independence under the Vasa dynasty in the early 16th century. After a long period of war and conflict, more stable and permanent frontier lines were established between Sweden and the twin kingdoms of Denmark–Norway from the beginning of the 18th century. In Sweden there was a fluctuating and uneasy balance built into the constitution between the Crown and the privileged Estate of the Nobility, while absolute monarchy was introduced in Denmark–Norway from 1660–61.[3] The Swedish nobility retained their predominance throughout the 18th century, while the power of the traditional Danish nobles was substantially diminished. In Denmark–Norway the rise of a new semi-official nobility and the emergence of a commercial bourgeoisie was favoured by the rather benevolent despotism of the monarchical rulers. Danish civil servants formed the new aristocracy in Norway. The twin kingdoms (i.e. a personal union under one king) were also, to some extent, economically interdependent. Most important was the Danish monopoly of corn import in southern Norway,

[1] Cf. Andreas Holmsen, *Norges Historie,* 3rd. ed., Universitetsforlaget, Oslo, 1961, pp. 354ff.

[2] Cf. ibid., pp. 396 ff.

[3] Cf. Nils Andrén et al, *Nordisk statskunskap,* Liber, Sth. 1963, pp. 25–29, pp. 125–126, pp. 156–158. For the background of the new central administration see the analysis by Knut Mykland, *Skiftet i forvaltningsordningen i Danmark og Norge,* Universitetsforlaget, Oslo 1973.

in accordance with the demands of Danish landowners, as foreign grain was excluded from September 1735.[4]

In the peace treaty of Kiel in 1814, after the Napoleonic wars, Norway was surrendered to Sweden. Sweden had by then joined the victorious coalition, while circumstances had led Denmark to an alliance with France against Britain. An extraordinary national assembly declared Norway's independence and signed the present constitution in May 1814, but as the union under the Swedish king proved inevitable, Norway gained only partial self-determination.[5] The union was finally dissolved in 1905. When Sweden acquired Norway in 1814, it was in compensation for the loss of Finland to Alexander I of Russia five years earlier. Finland had by then been united to Sweden for about seven centuries, and finally gained her independence after eleven decades of Russian overlordship by the Revolution of 1917.

Denmark is by far the smallest of the Scandinavian countries, with 16 575 square miles within her present borders, while Sweden has 175 350 square miles and Norway 124 557 square miles–stretching more than 1000 miles from south to north.[6] Substantial parts of the Scandinavian peninsula–Norway and Sweden–are situated north of the Arctic Circle: the northern cap of Sweden and most of the three long northern provinces of Norway. Because of the influence of the relatively warm Gulf Stream, the climate is substantially milder than elsewhere in the polar regions of the world. Western Norway has relatively chilly and rainy summers, and rather mild and stormy winters, while the summers in eastern Norway and most of Sweden are drier and warmer, and the winters colder, with more snow. The climate of Denmark and the southern districts of Sweden is more continental, with relatively warm summers and quite mild and snow-free winters.

The Danish flatland, especially in Zealand, the islands including Funen, and the eastern provinces of Jutland, is fertile agricultural land, and so are the southern provinces of Sweden. Norway is an extremely mountainous country, with deep fjords and steep valleys along the coast, and mountain plateaus and meagre highland districts inland. There are some areas of relatively fertile farmland in the east, around the Oslo fjord and the lake of Mjøsa, and in the central district of Trøndelag. The great eastern valleys are heavily wooded, and so

[4] Norwegian iron and glass were, on the other hand, given preferential treatment in Denmark.

[5] One of the most extensive discussions of these events is Sverre Steen, Det frie Norge, vol. 1, *1814*, Cappelen, Oslo 1952.

[6] *Longmans Dictionary of Geography* (ed. by Sir Dudley Stamp), Longmans, London 1966. There have been a lot of minor revisions of the border between Sweden and Norway, Sweden and Denmark, and Denmark and Germany thoroughout the centuries. In the 18th century Denmark had some areas which now belong to southern Sweden and northern Germany.

are the central and northern provinces of Sweden, with scattered patches of meagre farmland cleared along lakes and rivers.[7] In the 18th and 19th centuries the main arable area in Norway was thus the south-east and Trøndelag. The broad eastern valleys were forest area, while the central rocky valleys from the southern coast and northwards were pastoral land. The long coastal area in the west and the north was primarily dependent upon fishing. The plains and valleys of the far north were too arctic, with too short summers, to be usable for agricultural production. The nomadic Lapps have traditionally based their subsistence on seasonal reindeer herding, and will not be included in the present investigation.[8]

Population[9]

The following description of the Scandinavian population in the late 18th and early 19th century will present a short comparative survey of the development of total figures, urban vs. rural residence, distribution of occupations and social strata, and the variation in density of settlement within the geographical area. This synopsis aims at suggesting the character of the Scandinavian peasant societies in demographical terms, and also to serve as a general frame of reference for later sections of the study.

The total population of Norway, Denmark and Sweden by the middle of the 18th century was slightly more than 3.2 mill., or about 2 % of

[7] Cf. for instance the comparative description in Roy Millward *Scandinavian Lands,* Macmillan, London 1969.

[8] Per Otnes, *Den samiske nasjon,* Pax, Oslo 1970, gives a comprehensive account of the the Norwegian Lapp society.

[9] The primary sources of 18th and early 19th century population statistics should be treated carefully. They often employed different criteria for occupational groups from census to census, certain groups might be counted twice (i.e., if a peasant farmer was also a merchant, or a crofter was also a craftsman or even a 'pauper'), cf. Steen, op.cit., *Det gamle samfunn,* Cappelen, Oslo 1957, pp. 55 ff. Corrected secondary accounts are thus often more realiable than the original census register. The most extensive analysis of Norwegian population statistics is Michael Drake, *Population and Society in Norway 1735–1865,* Cambridge University Press 1969.
The Swedish census material is presented and analysed by Gustav Sundbärg, *Bevölkerungsstatistik Schwedens 1750–1900,* Sth. 1907, and in later contributions by Eli Heckscher and Gustav Utterström; see Heckscher, *Sveriges ekonomiska historia från Gustav Vasa,* I–II, and *Ekonomisk-historiska studier* (1963), and Utterström 'Population problems in preindustrial Sweden', *SEHR,* vol. 2, no. 2, 1954, and 'Population and Agriculture in Sweden ca. 1700–1830', *SEHR,* IX 1961. The Danish material is analysed by Aksel Lassen, in *Fald og Fremgang. Træk af befolkningsudviklingen i Danmark 1645–1960,* Universitetsforlaget i Aarhus 1965, and 'The Population of Denmark, 1660–1960', *SEHR,* XIV, 1966. (SEHR = *Scandinavian Economic History Review).*

the population of Europe.[10] Sweden accounted for more than 1.7 mill. of this number, while the Danes were not considerably more numerous than the Norwegians. The gap between Norway and Denmark also diminished towards the end of the century, since the population increase of Sweden and Norway nearly doubled that of Denmark.[11] During the 19th century, the annual rate of population growth in Scandinavia runs as follows:

	(Annual rate of growth per 1000)					
	1801–1820	1821–1830	1831 1840	1841–1850	1851–1860	1861–1870
Sweden	4.8	11.2	8.4	10,8	10.3	7.7
Norway	5.1	14.7	9.7	11.7	14.0	7.9
Denmark	8.3	9.3	7.2	9.8	12.3	10.4

Source: G. Sui. lbärg, *Apercus statistiques internationaux,* Stockholm 1960, p. 31; reproduced in D. V. Glass & E. Grebnik, 'World Population: 1800–1950', *The Cambridge Economic History of Europe,* Vol. VI, Cambridge University Press, 1965; and in Sima Lieberman, *The Industrialization of Norway 1800–1920,* Universitetsforlaget, Oslo, 1970, p. 27.

In *Norway* one estimate says that there were about 530 000 inhabitants in 1720.[12] There was an uneven increase throughout the century, with a marked decline during and immediately after the serious crop failures of 1740–42.[13] The following figures indicate the total Norwegian population from the middle of the 18th to the middle of the 19th century.[14]

1750 – c. 642 000	1810 – c. 896 000
1760 – c. 700 000	1820 – c. 968 000
1770 – c. 745 000	1830 – c. 1122 000
1780 – c. 770 000	1840 – c. 1240 000
1790 – c. 802 000	1850 – c. 1392 000
1800 – c. 881 000	1860 – c. 1595 000

10 Cf. Adolph Jensen, 'Befolkningsforhold i de nordiske Lande i det 18. Aarhundrede', *Nationaløkonomisk Tidsskrift,* Copenhagen, 1935, p. 13, and Sundbärg 1907, op.cit., pp. 2–3.

11 Jensen, op.cit., p. 13. Sweden got orderly conditions for censuses and population statistics as early as 1748, and Denmark–Norway from 1769.

12 Steen, *NFLH,* VI, pp. 152 ff. This number is estimated backwards from the specification of persons born and deceased made by the priests, and supplemented by certain other corrections.

13 The figures for each year from 1735 to 1865 are presented by M. Drake, op.cit., pp. 164–167. See also the discussions of population growth, with further references, in Mykland et al., *Norge under eneveldet 1660–1720,* Universitetsforlaget 1975, p. 89ff., and Dyrvik et al., *Norge under eneveldet 1720–1800,* 1976, p. 25ff.

14 These numbers are rounded off from Drake's figures, op.cit. Cf. his discussion of the presuppositions on p. 167.

This means an average annual population increase of 0.74 % from 1750 to 1800, and 1.25 % from 1800 to 1860.[15] During the whole period covered by this study, the Norwegian population was overwhelmingly rural. In the latter part of the 18th century and well into the 19th, more than 90 % of the population lived in the countryside. It is estimated from the census of 1801 that still less than 9 % were urban residents, and by the middle of the 19th century the proportion was only slightly more–still just above 12 % in 1845.[16]

A somewhat inaccurate summary of the 1769 census says that the peasantry numbered about 420 000, including wives and children, while there were about 120 000 servants, 21 000 'poor' and 90 000 sailors –which is slightly more than 650 000 'non-bourgeois' subjects in all.[17] On the other hand, the bourgeoisie constituted less than 48 000, with about 35 000 of the commercial class and 13 000 connected to the central and provincial officialdom.[18] Still in 1835 more than 4/5 of the population were directly connected with agriculture.[19] From the census of 1801 it is estimated that the peasant society consisted of about 83 000 independent farmers (free tenants and peasant proprietors), and 203 000 crofters, day labourers and rural servants.[20] In 1835, there were about 103 000 peasant farmers and 225 000 rural 'dependents', and towards the middle of the 19th century there was a particular increase in the proportion of crofters, as we shall see later.[21] Even for the earlier period, from 1723 to 1825, it is estimated that the number of peasant farmers increased by about 33 % (with a maximum of 65 % in some eastern provinces), while the number of crofters with plots of land increased by more than 300 %.[22]

Norway also had a very scattered population. Vast areas were completely unpopulated, while the average density was about 2.85 inhabitants per km^2 in 1801 and 3.86 in 1835.[23] Most of the people were concentrated in the areas with arable land or along the coastal strip. The vast northern county of Finnmark had for instance less than 1 % of the total rural population in the second half of the 18th century,

[15] The growth accelerated from about 1820 onwards, as indicated by the table above.

[16] See Keilhau, *NFLH*, VIII; and Steen, op.cit., *Det gamle samfunn*, p. 20.

[17] Steen, *NFLH*, VI, p. 300. About 20 000 of the population, soldiers etc., are excluded from the figures.

[18] Ibid. The 'bourgeoisie' include mainly licenced merchants, guild craftsmen and civil servants, with their families.

[19] Steen, op.cit., *Det gamle samfunn*, p. 24.

[20] Ibid., p. 56.

[21] Ibid., p. 59.

[22] Cf. T. H. Aschehoug, 'Statistiske studier over folkemængde og jordbrug i Norges landdistrikter i det 17de og 18de aarhundrede', *Statsøkonomisk Tidsskrift*, Kristiania, 1890. Tillæggshefte. Aschehoug 1890. A table, with some reservations as to the reliability of the estimate, is presented on p. 38.

[23] Cf. Steen, op.cit., *Det gamle samfunn*, p. 20.

while the substantially smaller southern counties generally had about 5–6 % each, with particularly strong concentrations around the bigger cities.[24]

The population in *Sweden* by the middle of the 18th century was slightly more than 1³/₄ mill. [25] The number passed 2 mill. just before 1770, it was about 2¹/₃ mill. in 1800, then passed 3 mill. before 1835, and 4 mill. at the beginning of the 1860s.[26] Even if there was a steady increase from decade to decade, the variations from one year to another could be quite big–especially during the 18th and early 19th centuries, when war, pestilence and crop failures heavily influenced marriage patterns, birth-rates and mortality.[27] In comparison to the above figures for Norwegian population growth, we might cite Sundbärg's classical estimates:[28]

1750 – c. 1780 000	1810 – c. 2396 000
1760 – c. 1925 000	1820 – c. 2585 000
1770 – c. 2042 000	1830 – c. 2888 000
1780 – c. 2118 000	1840 – c. 3139 000
1790 – c. 2188 000	1850 – c. 3482 000
1800 – c. 2347 000	1860 – c. 3860 000

This is an average annual increase of 0.64 % from 1750 to 1800, and 1.07 % from 1800 to 1860.

Sweden was hardly more urbanized than Norway. Slightly more than 9 % of the population were urban dwellers in 1760, and in 1800 the proportion was still less than 10 %.[29]

But contrary to the situation in Norway one city, Stockholm, was completely dominant among Swedish towns, and constituted nearly 40 % of the urban population, with about 70 000 inhabitants, by the middle of the 18th century.[30] Stockholm had grown by more than 50 % in the 40 years before the late 1750s, but then the growth stagnated,

24 Aschehoug 1890, op.cit., table on p. 14.

25 Cf. Sundbärg 1907, p. 75; and Heckscher, *Svenskt arbete och liv,* Bonniers, Stockholm 1957, p. 150.

26 See Sundbärg, 1907, p. 75, with a critique of sources on pp. 161 ff. Sundbärg's figures differ somewhat from those of Heckscher 1957, p. 150, which again differ from those of Utterström, *SEHR,* 1954, op.cit. But the difference is not substantial.

27 There was recently a debate about the impact of these factors on Swedish population growth–cf. Utterstöm, *SEHR,* 1954, pp. 101–165, and Bertil Boethius, 'New Light on Eighteenth Century Sweden', *SEHR,* pp. 150 ff. Utterström polemizes against the 'Malthusianism' of Heckscher's *Sveriges economiska historie från Gustav Vasa,* which is defended by Boethius.

28 Sundbärg 1907, p. 75. Here the figures are rounded off.

29 Cf. Heckscher 1957, op.cit., p. 161.

30 Ibid., p. 162.

and the population of the city was still only about 75 000 in 1820, probably owing to commercial changes and to the declining immigration after the loss of Finland in 1809.[31] Stockholm had about 1/3 of the urban population in 1800, and about 1/4 in 1850, as the process of urbanization spread to other local centres.

The agrarian population in Sweden proper (excluding Finland) was about 3/4 of the total population in the second half of the 18th century, and this substantial proportion stayed nearly unchanged over the next hundred years–in 1870 the proportion was still more than 72 %.[32]

The dominant nobility constituted only 0.5 % of the population in 1760, the clergy and the non-noble 'persons of rank' about 2.5 %, the urban commercial class 6 %, and lower official servants–mostly soldiers–about 9 %. Then, on the other hand, the real peasantry constituted nearly half of the population, and crofters and rural lodgers more than 10 % each.

Geographically, the density of settlement varied greatly.[33] The vast northern provinces were nearly uninhabited by the middle of the 18th century, with less than 1 inhabitant per km^2 on the average. The density was highest in the south and south-east, with 15–25 persons per km^2, against 5–15 over the whole southern half of the country. Not surprisingly, this distribution corresponded closely to the quality of arable land.

The ecological conditions of *Denmark* differ markedly from the heterogeneous situation on the Scandinavian peninsula, and so do the development and distribution of the demographical pattern. According to one estimate Denmark had about 806 000 inhabitants in 1750.[34] Then the population growth runs as follows:[35]

1770 – c. 810 000	1840 – c. 1280 000	1860 – c. 1608 000
1800 – c. 935 000	1850 – c. 1415 000	

This is an average annual increase of 0.32 % from 1750 to 1800, and 1.20 % from 1800 to 1860.

Denmark had a much greater proportion of urban dwellers than Norway and Sweden. As early as 1769 nearly 21 % of the population lived in towns, and Copenhagen alone, with nearly 83 000 inhabitants, had about 51 % of the urban population.[35] The population of Copen-

[31] Ibid., p. 162.

[32] Ibid., pp. 158–159.

[33] Ibid., p. 155.

[34] A. Jensen 1935, op.cit., p. 13.

[35] A. Lassen 1965, op.cit., tables on pp. 540–541. The main reservation here is that the Danish area probably is not exactly the same as it was for Jensen above. Cf. Lassen *SEHR*, XIV, 1966, pp. 134–157, who gives corrected figures for Denmark excluding the province of Slesvig, and who turns up with a slightly different estimate of the population in the year 1800.

[35] Cf. Lassen 1965, tables with a comment on the estimates on p. 365.

hagen passed 90 000 in 1787, when the other Danish cities still had slightly more than 82 000 taken together, and in the year 1800 it passed 100 000 inhabitants.[36] Thus Danish agriculture produced a surplus to feed a substantial urban population at a time when Sweden, and particularly Norway, was still hardly above the rural self-subsistence level.

In 1769 the so-called 'persons of rank'–noblemen and higher civil and military officials–constituted 1.7 % of the total population (more than 5 % in the cities and less than 1 % on the countryside), while the clergy amounted to nearly 2 %.[37] Industrialists, craftsmen and commercials made up for 14 %. Servants and day labourers constituted nearly 19 %, while the peasants and crofters (including children) amounted to about 57 % of the population. The Danish society too was heavily based on a numerous peasantry, even if these were related to a more clearly visible non-peasant urban world than their counterparts further north.

Denmark was also much more densely populated than the Scandinavian peninsula. The counties of north-eastern Zealand, including Copenhagen, had a density of more than 60 inhabitants per km^2 in 1769, and nearly 100 in 1840.[38] The island of Funen had 31 per km^2 in 1769 and 53 in 1840, while western Jutland correspondingly increased from 9 to 13 in that period, and eastern Jutland from 18 to 30. Nearly half of the population lived on Jutland during the whole period, while Zealand definitely was most densely populated. This pattern also corresponds neatly to the quality of climate and arable land, and so does the over-all distribution of settlement within Scandinavia as a whole.

Land tenure systems

By land tenure systems we mean both the distribution and type of landed property, and the organization of holdings within farms and villages. We will first make a comparative survey of landed property relations in the three Scandinavian countries prior to the agrarian transformation from the end of the 18th century. Then the different features of 'farm structure' will be compared.

The following factors make strict cross-national comparisons rather complicated:

Firstly, there are great regional variations not only within the Scandinavian area as a whole, but within each of the three countries themselves. The geographical conditions vary least in Denmark, but are highly heterogeneous in Sweden, and extremely so in Norway. How-

[36] Ibid., tables on p. 366 and p. 372.
[37] Ibid., table on p. 364.
[38] Ibid., table on p. 558.

ever, the regional factor will especially be spelled out in a closer investigation of Norwegian regions. There are also certain basic national characteristics due to separate legal institutions, taxation decrees and historical events which make the nation-state a meaningful unit.

Secondly, we face an unfortunate insufficiency of corresponding figures. The source material does not give quite equivalent units and categories, and exact calculations are hard to achieve. Our aim is then limited to giving approximate indications of significant patterns.

Thirdly, inter alia, the category of 'freehold' has different implications in the various settings, both *as such* and in relation to alternative types of landholding. 'Peasant proprietorship' implies different qualities in each of the national polities, as well as a multiple content within each country, both synchronously and over time.

Fourthly, there is no parallel historical development in the Scandinavian area as a whole. There are, on the contrary, different processes taking place, with dissimilar sequences and dissimilar chronology and time-order.

The analytical implications of these factors will be made clear during the following discussion.

Peasant freehold

On the whole, the share of the land possessed by the Scandinavian peasantry had decreased during the 16th and early 17th centuries.[39] It is generally assumed that a decline of peasant freehold took place in Norway from the Middle Ages, although peasant property diminished less than in Sweden and Denmark, and never reached as low a proportion as in these countries. In Danish history the period up to 1660 is usually called 'The Age of the Nobles', and according to *Bolin* about 10 % of the land was peasant property in the middle of the 17th century.[40] Another estimate says that about 6 % of the Danish peasantry were freeholders at that time (1650), while the amount had been double that number a hundred years earlicr.[41] The freeholders were as a whole an insignificant class. Shortly before 1700 there were still less then 1 % freeholders in most parts of the country, with at most 3 to 8 % of the farmers owning their land in a few counties.[42] The rest were

[39] Cf. Sture Bolin, 'Medieval Agrarian Society in its Prime: Scandinavia', in M. M. Postan (ed.), *The Cambridge Economic History of Europe,* Vol. 1, 2 ed. Cambridge University Press, 1966. Chapter VII, 8, pp. 633–659.

[40] Ibid., pp. 658–659.

[41] Cf. Huidtfeldt, 'Bondegard og Herregård', in *Den Danske Bondegård,* Odense 1965, p. 19. The estimate is elaborated by J. A. Fridericia, 'Historisk-statistiske Undersøgelser over Danmarks Landboforhold i det 17de Aarhundrede', (Danish) *Historisk Tidsskrift,* R.6, B.2, 1889–1890, pp. 469–622, who must be Huidtfelt's source.

[42] Cf. Fridlev Skrubbeltrang, 'Developments in Tenancy in Eighteenth Century Denmark as a Move towards Peasant Proprietorship', *SEHR,* IV, 1961, p. 165.

highly dependent leaseholders, and it should be noted that the freeholder had smaller duties–according to Christian V's 'Danish Law' of 1683 he had only 1/4 of the considerable labour service *(hoveri)* imposed on the tenant.[43] The duties will be specified below.

In Sweden the taxpaying peasants (i. e., freeholders paying taxes to the State) possessed about half of the registered land by the middle of the 16th century, but less than 1/3 about 1700. The rest was fairly equally divided between the Crown and the taxfree nobles.[44]

There were also regional variations in Denmark and Sweden. Denmark as a whole was more fertile, with more arable land and fairly uniform conditions of nature and settlement, compared to Norway and Sweden. Thus Denmark had fewer regional variations, even if the slightly less fertile Jutland (especially the western area) also showed the greatest tendency towards freehold. It is also striking that the greatest proportion of peasant freehold in Sweden was concentrated in Finland and Norrland, both colonized late and being poor peripheries. Here, as in the district of Dalecarlia, a good majority of the peasant population was landowning farmers by the middle of the 16th century, while 'it is typical for the general situation that in the finest agricultural district of Sweden, Oestergoetland, only a seventh of the peasantry owned their land'.[45]

We will come back to a more detailed survey of the character of Danish and Swedish freeholds below, but first a closer investigation of the Norwegian stituation.

The great transition towards peasant proprietorship in Norway took place extremely early–the first wave already flowing from the sale of crown property in the last third of the 17th century. The main characteristics of this development will be discussed in the next chapter. Here we will sketch the structural patterns of land ownership prior to the great processes of change. It was a tradition in Norwegian historiography in the past to give rather exact figures of the distribution of land between tenants and 'yeomen' freeholders, between *leilendinger* and

[43] Ibid., pp. 165–166.

[44] See Heckscher 1957, op.cit., p. 144. The main change since the Middle Ages was that the church estates had been seized by the King.

[45] Bolin, op.cit., p. 659. On Norrland as an open frontier for the peasantry, cf. Elfred Kumm, *Jordhunger och dagsverkstorp,* Sth. 1949, p. 7.

[46] Cf. Ernst Sars, *Udsigt over den norske Historie,* vol. I–IV, Kristiania 1873–91, on this special point see *Samlede Værker* II, Kristiania and Copenhagen 1912, p. 284, pp. 301–307. O. A. Johnsen, 'Fra leilending til selveier', *Nordisk Tidsskrift,* 1910.
Simen Skappel, 'Det norske jordleievesens geografiske utbredelse i eldre tid og overgangen fra leie til selveie', *Historisk Tidsskrift,* vol. 30, Oslo 1934–36.
Ivar Mortensson, *Bondeskipnad i Norig i eldre tid,* Olaf Norli, Kristiania, 1904, p. 10.
Halvdan Koht, *Norsk bondereising,* Aschehoug, Oslo 1926, p. 204.
Sigvald Hasund, *Vårt landbruks historie,* Aschehoug, Oslo 1932, p. 96.
Steen, *NFLH,* vol. V, pp. 262 ff.

odelsbønder, prior to the extensive sale of Crown land after 1660.[46] *Sars* and his successors estimated that 1/4 of the peasants were freeholders and 3/4 were tenants by the middle of the 17th century. The regional variations were emphasized. Tenancy dominated in the western coastal districts of the country, and also in the fertile corn and timber districts in the east. *Koht* estimates the relation of freeholders to tenants as about 1 to 5 in two central districts in the east, while the two groups were evenly distributed in the valley of Gudbrandsdalen.[47]

The clear distinction between tenant and freeholder has been questioned by later research. But it served two obvious functions besides the satisfaction of historical curiosity: first to give a clear picture of the Norwegian freeholder, the *odelsbonde,* as a genuine national hero, descending from the golden Nordic past; and then to show the clear development of peasant freehold during the 18th century, the restoration of traditional strength as a prerequisite for national independence and maturity later on. The distinctiveness of the traditional freeholder might thus have been retrospectively exaggerated as a justification for subsequent political movements.

However, the material reality of 'ownership' was not the same during the age of tenancy as it turned out to be in the 18th and 19th centuries; ownership was generally a rather complicated legal affair which could not easily be connected to specific pieces of land, and the tax rolls from which national historiography extracted its *odelsbønder* blurred many local varieties.[48] Owing to the specific system of land tenure in force by the middle of the 17th century, it is thus difficult to tell, with some pretention of meaningfulness, how many peasants were freeholding proprietors, or how many 'farms' the freeholders possessed relative to estates let on lease. What can be done, and what recently has been done for substantial parts of the country, is to estimate the distribution of total rent value, the *landskyld,* between different categories of owners.[49]

A fairly well-founded estimate indicates that slightly less than 1/3 of total assessed rent value was peasant property when absolute monarchy

47 Koht 1926, op.cit., p. 204.

48 See Andreas Holmsen, 'From tenancy to freehold peasant ownership', *SEHR,* IX, 1961; Halvard Bjørkvik, *Jordeige og jord-leige i Ryfylke i eldre tid,* Stavanger 1958, pp. 60 f.

49 The *landskyld* was originally a leasehold rent which was paid in kind in a fixed proportion to the yields of the farm. Later it was attached to *all* farms as a measure of their value and a basis for taxes to the State and division of inheritance. All farms were assessed *as if* they were tenant property, and there were special norms for estimation which were proportional to seed, livestock and 'accompanying amenities'. The *landskyld* as a measure of use value and a basis for tax assessment stimulated more complete land registration from the middle of the 17th century.

was established in1661.[50] But the regional and geographical variations are really striking.[51] In the north, the centre (Trøndelag) and the coastal districts of the west and south-west, peasant property was relatively smallest–about 1/10 of the *landskyld*. Both in the eastern flatland region and in the inner fjord and mountain districts of western Norway peasant property amounted to about 1/3 of the rent, while in a belt from the south towards the central parts, mainly composed of the inner mountain districts of eastern Norway, peasant freehold was absolutely dominant, with more than 4/5 of total rent value. Peasant freehold was thus negligible along the western coast and northwards, where the main means of livelihood were derived from the sea rather than from agriculture, and it was dominant in the peripheral and poor mountain regions. The central agricultural districts in the east were in an intermediate position.

This pattern seems to be confirmed within the regions and local areas themselves.[52] In a district of south-western Norway, for instance, the outer and central communities were characterized by very little peasant property, while the proportion of freehold increased the more interiorly and remotely the communities were situated.[53] The landowning peasantry had their strongest position in the mountain area and the narrow interior of the fjord districts, without large fjords and islands, and without great riches of any kind.[54] The Crown, the church, or the landed nobles had not expropriated the freeholding peasantry in the interior districts, which were mainly based upon livestock husbandry and farming, but did do so along the coastline, where riches were extracted from fishing, whaling, and sealing, and which offered strategical control of the main fairway northwards.[55] Even in the agricultural plain and fertile valley area of eastern Norway, the peasantry possessed a substantial share of the land at some distance from the commercial and administrative centres (i.e. mainly Oslo), while *burgher* property and the estates of civil servants were concentrated in sawmill districts

50 See Holmsen, *SEHR*, 1961, op.cit., p. 162; Holmsen, *Gard, Bygd, Rike*, Universitetsforlaget, Oslo 1966, pp. 233–234; Holmsen & Bjørkvik, 'Hvem eide jorda i den gamle leilendingstida?', *Heimen*, vol. IX (1952–54). These works are based on source material from the land registration of the ordinary land commission of 1661.

51 Cf. the tables compiled by Holmsen 1961, op.cit., p. 162; Bjørkvik 1958, op.cit., pp. 56–57; Holmsen & Bjørkvik 1952–54, op.cit.

52 Bjørkvik 1958, op.cit., Holmsen & Bjørkvik 1952–54, op.cit., Holmsen 1966, op.cit. pp. 54 ff.

53 Cf. Bjørkvik 1958, p. 114, note 9.

54 Cf. Holmsen, 'Landowners and Tenants in Norway', *SEHR* VI, 1958, p. 128.

55 See Holmsen ibid., and Bjørkvik 1958, pp. 74 ff. In the age of national unification there had been a violent antagonism between royal power and old chieftain dominance along the coastal fairway, while the simple peasantry dominated in rural districts which were cut of from the events cn the main stage of the kingdom.

and along the central rivers for timber rafting, stimulated by the commercial significance of the lumber trade in the first half of the 17th century.[56] Generally, land as such does not seem to have been especially attractive to the upper classes at this stage of agricultural development.

We have seen that peasant property is estimated as percentages of total assessed rent value, the *landskyld*. This is not an indirect indicator of the degree of freehold but an exact expression of the meaning and implication of peasant proprietorship.[57] The relations of ownership and land use were extremely complicated in 17th-century Norway. Very often there were many owners to each registered farm, and this shared ownership constituted a right to certain parts of the annual rent value, the *landskyld*, not, or only occasionally, to specific pieces of land or marked fields and meadows. Thus the *landskyld* of an individual farm could be shared in fixed proportions between the state, the church, a nobleman, a civil servant, a commercial city dweller, and one or more peasants.

Among the different owners the one who possessed the largest part had the 'lease right' (*bygselsrett*), i.e. the privilege to decide which tenant should be given the farm on lease and to reserve the entry fine and other lease payments, outside the annual rent, for himself. If two owners possessed equal parts, the one of higher rank got the lease rights. Now a peasant farmer could own a part of the land he used, but might pay the annual rent for tenancy to the owners of the remaining parts of the farm. If his own part was not the largest, he also had to pay lease fines to the dominant rent-proprietor. Correspondingly, a peasant could be a tenant in relation to the farm he and his family worked, but an owner of parts of one or more farms used by others. Thus the distinction between tenants and freeholders could be extremely obscure.

This indistinctiveness was accentuated by the inheritance system. The oldest son had the legal right to take over the farm without partition,[58] but if he could not buy out his co-inheritors, as was usually the case, he used their parts on lease and paid them the annual rent as a tenant. To be a really independent freeholder, an *odelsbonde* of the slightly mythical Old Norse type, then, the peasant not only had to be an owner of land, but to be the *only* owner, or to control the leasehold, of the land he used. Outside the remote and meagre areas of southern Norway, the districts of Agder and the interior of Telemark, this type

[56] Cf. Holmsen 1966, op.cit., pp. 54 ff.

[57] See Holmsen 1958, op.cit.; Holmsen 1966, op.cit., pp. 123–134; Bjørkvik 1958, op.cit.; Bjørkvik, 'Leilending', *Kulturhistorisk Leksikon for Nordisk Middelalder,* Gyldendal, Oslo 1965, X, p. 462; Steen *NFLH,* IV, pp. 205 ff.

[58] Through the *åsetesrett*, a legal institution whose origin is discussed by Holmsen, 'Om opphavet til åseteretten', in Holmsen 1966, op.cit., pp. 110–114.

was rather rare.[59] Generally, the age of tenancy before the 18th century was not characterized by individualized separate estate or distinct rights of property and usage. The *landskyld* was a segregator between proprietorship and land.

However, even if the distinction between freehold and tenancy was somewhat obscure in many Norwegian areas, a substantial proportion of peasant proprietorship was real enough. This was not unreservedly the case in Sweden, and it was by no means the case in Denmark.

Until the decades around 1800 peasant freehold in Denmark was quite negligible. There were, as we have noted, a few percentages of peasant proprietorship in certain regions, mainly in the duchies, on the island of Bornholm and some districts in western Jutland. But even this was a freehold within much more restrictive limits than the prevailing Norwegian type. The rent of the freehold was split into 'peasant rent' (*bondeskyld*) and accompanying amenities (*herlighed*).[60] The owners of these amenities, the Crown or a noble or bourgeois estate-owner, could claim certain rights and privileges at the expense of the peasant. They had economic rights, like a charge on the peasant's produce, annually or once and for all; a right to labour service (although only a fourth of the duties of tenants); and the privilege of fishing and hunting on the peasant's land. Further, the peasants were subject to patrimonial domain even as 'freeholders'. The landed lords were given limited jurisdiction within their district, the so-called *birkerett* and *sagefald*,[61] and a privilege of legal guardianship over the peasantry.[62] The *jorddrott*, the lord of the land, was a person of rank and a formal mediator between public authorities and the simple peasant. Thus the destiny of the freeholding peasant was not particularly different from that of the *faestebonde*, the dependent tenant. His duties were more restricted, but proprietorship was limited within the general relations of patrimonial domain.[63]

59 Cf. Holmsen 1966, tables on pp. 233–234, and Holmsen 1961, p. 160. The ethnologists tell us that special local traditions are confined to the area of old peasant proprietorship and expressed in the legend of the *deildegast*– a restless ghost who haunted the fields in pain because he, while alive, had removed the boundary stones and stolen land; cf. Johannes Skar, 'Gamalt or Setesdal' II, Kristiania 1907, p. 152, ref. by Gudmund Sandvik, 'Jordleigar og jordeigar i norsk historie', stencil 1969, p. 1.

60 The partition of the total amount of *hartkorn*, the Danish unit for measurement of the land rent, into a double assessment, was made in the land register of Christian V.

61 Cf. *Kulturhistorisk Leksikon for Nordisk Middelalder*, 'Bjärköarett', vol. I, Gyldendal. Oslo 1956, pp. 655–661, 'Sagefald', vol. XIV, Gyldendal, Oslo 1969, pp. 651–657.

62 Cf. Sigurd Jensen, *Fra patriarkalisme til pengeøkonomi*, Gyldendal, Copenhagen 1950, pp. 15 f.

63 The Weberian notion of patrimonial domain implies that control over the peasantry is inherited by landlords who have a right to receive tribute from the personally dependent peasants in return for their land occupancy; cf. Eric Wolf, *Peasants*, 1966, pp. 50 f.

In Sweden, as we have seen, the peasantry possessed about 1/3 of registered land at the beginning of the 18th century, varying regionally from a clear majority down to the Danish average. But these Swedish *skattehemman* ('taxable farms') do not have the same institutional meaning as the corresponding Norwegian *odelsgard* ('allodial peasant farm'), even if both might be translated as 'freeholds'. Swedish land was originally divided into *skattehemman, kronohemman* (crown estates) and *frälsehemman* (the 'free' land of the nobility). The nobility possessed taxfree land, with more or less complete proprietorship, originally as a compensation for military service to the king. But ownership of peasant land was legally restricted, not for the benefit of the lords of the land, as was principally the case in Denmark, but directly for the benefit of the Crown. To the crown land the king had both the right of use (lat. *dominium jutile*) and the right to the yields of the soil (*dominium directum*). The ownership of the peasant land, the *skattehemman,* was both legally and in actual fact divided, with the right of use belonging to the peasants, but the *dominium directum* being the real property of the Crown. Thus the rent claimed by the king for this actual domain was not regarded as a public tax, but as a private income from his own property. Accordingly, there were also strict regulations on the disposition of peasant land. Peasant freehold in 17th and 18th century Sweden should thus be interpreted within the general framework of feudal notions of land ownership.[64] A substantial increase in peasant land had taken place at the expense of the nobility in the last decades of the 17th century. This was the so-called 'reduction' (*reduktionen*), led by the king, to support the state financially by transferring taxfree land, which had once belonged to the Crown, back to the taxpaying peasantry.[65] But this affected scattered farms and pieces of land, the *strøgods,* from which the nobility had drawn rents, rather than the home farms and manors themselves, and was no hard blow against the position of the landed nobility.[66] After the end of the 'reduction', about 1700, the nobles still possessed 1/3 of Swedish land,

[64] Cf. Eino Jutikkala, *Bonden i Finland genom tiderna,* LT's Förlag, Helsingfors 1963. pp. 154 ff.
See also Lars Herlitz, *Jordegendom och ränta*, Lund 1974.

[65] Already in the late Middle Ages legislation was introduced to check transactions whereby peasants sold their land to taxfree nobles and afterwards cultivated it as tenants–which would also have caused the income of the Crown to decline. See Bolin, op.cit., p. 658. The explicit notions of feudal land ownership were introduced as a later legitimation–cf. Jutikkala, op.cit., pp. 154 ff.

[66] Cf. Heckscher 1957, op.cit., pp. 135–145. Former *kronobönder* ('crown peasants') who had become *frälsebönder* (peasants on noble land) by donations, also became *kronobönder* again; and former *skattebönder* ('tax-paying peasants') who had become *frälsebönder* by donations, became *skattebönder* once more – cf. Ingers, *Bonden i svensk historia,* (3 vols. Stockholm 1943–56), vol. I, pp. 284 ff.

and the peasantry had another third at their partial disposition, whereas their share at the time of the death of king Gustav Vasa, in 1560, had been more than half.[67]

Landed estate

Before the transition towards freehold in Norway, beginning with the sale of crown land after the introduction of absolute monarchy in 1661, about 70 % of the total rent was owned by non-peasants.[68] This was crown property, church property, noblemen's property, and burgher property. After the confiscation of church estates during the Reformation, the king had become the biggest landowner in the country. By the middle of the 17th century, crown and fief estates amounted to nearly 1/3 of the *landskyld,* but there were important regional variations. In the central district of Trøndelag, with much church land formerly concentrated around the cathedral in Nidaros in the period preceding the Reformation, the king possessed about half of the rent.[69] Crown property was also dominant in northern Norway with about 40 %, and the same along the western coast. In the eastern plain and fertile valley area between 1/4 and 1/3 was crown land, while the king possessed only about 1/10 of the land in the strongholds of peasant freehold in the mountain regions of the south and central east.[70] Church property was strongly concentrated in the north and in the county of Trøndelag, with about 1/3 of the land, and between 1/4 and 1/5 both in the east and the west. But in the southern and eastern mountain area, again, ecclesiastical property did not surpass 1/10 of the rent.

Noblemen's property was extremely modest in Norway. Most of it was to be found in the north and along the western coast, but even here amounting to only 1/10 of the land. Then there was a small concentration in some areas south-west of Oslo, but for the remaining parts of the country the landholding nobility was insignificant.[71]

Geographically, this pattern is the inverse of the distribution of freeholds. The landed property of the central powers taken together, i.e. the land of the king, his servants and creditors, and the central ecclesiastical institutions,[72] was dominant along the coast, on the islands and close to the administrative centres, secular or sacred. The strategical sites on the fairway northwards were, once in history, controlled by

[67] Ingers, op.cit., p. 292.

[68] Cf. Holmsen 1961, op.cit., and 1966, op.cit.

[69] Holmsen's figure for 1661 (1966, op.cit., p. 233), is 55 %, while Bjørkvik's estimate for ca. 1650 (1958, op.cit., p. 117) is 46 %. Cf. *Heimen* IX (1952–54).

[70] Ibid.

[71] Cf. Holmsen 1966, op.cit., pp. 233–234.

[72] By central ecclesiastical institutions we include sees, chapters, cathedrals, and monasteries, but we exclude the land of the village priests and their churches.

local chiefs, and had to be conquered by the central power during the period of national unification and later.[73] However, after the political centralization of Denmark–Norway, by the introduction of absolute monarchy in 1660–61, landed property and local position turned out to be less important–and financial needs more so–as a basis for political power. The later sale of crown land could thus be seen as an indicator of the relatively limited importance of agriculture as a source of wealth.[73a] This view is supported by the increase in bourgeois landed proprietorship during the 17th century. This was mainly stimulated by the growing timber commerce rather than by agriculture itself, and bourgeois property was often concentrated in sawmill districts and along the transport rivers in the east.[74] Similar processes took place in the fishing communes along the western coast, at the expense of both the former nobility and the peasant freeholds. This economic inducement affected not only the commercial urban interests, but also the local civil servants, who possessed farm-land as a part of their public income.

There were mainly two types of management of landed estates in Norway. The dominant type was tenancy. The ownership of land often consisted of scattered parts of the rent of registered farms within a wide geographical area. The system of rent tenancy was based on a free contractual relationship between the proprietors and the tenant farmer. Even the landed nobility possessed substantial parts of its land as scattered leasehold property, probably making Norwegian estates much less attractive to agriculturally minded noblemen than the considerable amount of manors and demesne-land in Denmark.[75] The other type was unified manor farming, *setegårdsdrift,* both by bourgeois proprietors sitting on a home farm, and crown and church fiefs used by civil servants or noblemen. These home farms were worked by hired servants–only occasionally supplemented by a limited amount of duties from the peasantry living within a certain circumference.

To sum up, there are thus three notable qualitative peculiarities, intrinsic to the structure of landed estates in Norway, which can be traced back to the 17th century and further:

Firstly, the system of free, contractual rent tenancy.

Secondly, the dominant composition of the estates as scattered parts

[73] Cf. Holmsen, *Norges Historie,* 1961, op.cit., Bjørkvik 1958, op.cit., pp. 74 ff.

[73a] See the qualification in Østerud, 'Configurations of Scandinavian Absolutism', paper to ISSL/MSH Symposium, Bellagio, October 1976.

[74] Cf. Holmsen, 'Fordelingen av jordeiendommer i Oslos opland før krongodssalget' in *Gard, Bygd, Rike,* 1966, op.cit., pp. 54–72.

[75] Cf. Bjørkvik 1958, op.cit., tables for Rogaland/Agder and Bergenhus counties in 1661, pp. 56–57; Holmsen, *SEHR* 1958, op.cit., p. 130.

of the rent of different farms, i.e. the amount of *strøgods*, 'scattered property'.

Thirdly, the limited existence of a landed nobility.

These aspects of the Norwegian agrarian system build an interrelated whole, with the peasant farm at its base.

During the 17th century there was increasingly a transition towards large-scale manorial farming both in Sweden and Denmark. This concentration into unified estates (Dan. *hovedgårder,* Swed. *säterier*) has been connected with the general agrarian depression in 17th-century Europe, and seems quite parallel to the transition from *Grundherrschaft* to *Gutherrschaft* in Germany and Eastern Europe.[76] In Denmark the development of estates with large-scale farming had already started in the 16th century, when manors were separated from the communal ownership of land, and the development continued into the second half of the 18th century.[77] True, in a Danish law of 1683 it was prohibited to incorporate peasant farms into large-scale manors, but this prohibition never proved effective. Especially during the second half of the 17th century much peasant land and deserted farmsteads were annexed by the manorial home estates.[78] The incentive to exchange of real property, to concentrate the landed estates close to the home farm, was doubled: it was partly stimulated by the fact that the manorial land of the nobility was taxfree, as were the duties of the peasants living nearby the manor, the *ugedagsbønder* ('week day peasants'). And partly the process was stimulated by the growing possibilities of cheap labour service, both from the duties owed by an increasing number of peasants living within the domain of the extended home farms, and from peasants who were expropriated and declassed into cottagers on their former tenant farms. When the estates were concentrated into large-scale farming, villeinage (*hoveri*) could be increased so that the number of people and horses on the manors themselves could be reduced. Production for the market could thereby be stimulated at the expense of subsistence production. The duties also

[76] '*Grundherrschaft*' implied relatively restricted domain for the landlord, where duties from the tenants were mainly collected in rent. '*Gutherrschaft*' meant principally large-scale farming by compulsive labour duties from peasants who were hereditarily adscript to the manorial estate. The system of '*gutherrschaft*' was especially accentuated in Prussia and Poland from the middle of the 17th century–cf. Kåre Lunden, 'Agrartilhøve ca. 1600–1720', *Historisk Tidsskrift,* vol. 48, No. 4, 1969, p. 227, note 26, and pp. 230 ff.

[77] Cf. Gunnar Olsen, *Hovedgård og bondegård. Studier over stordriftens udvikling i Danmark i tiden 1525–1774,* Landbohistoriske Skrifter, vol. 1, Rosenkilde & Bagger, Copenhagen 1957, pp. 76 ff.

[78] Cf. Hvidtfeldt, op.cit., p. 24; Gunnar Olsen, 'Det unge Enevælde 1660–1721', John Danstrup & Hal Koch (eds.), *Danmarks Historie,* vol. 8, politikens Forlag, Copenhagen 1964, pp. 318 ff; Lunden 1969, op.cit., p. 230.

spared the manor the burden of agricultural tools as an item of expenditure. Briefly, manorial farming meant increasing incomes and reduced expenses. For the peasant farms, on the other hand, villeinage led to an increasing need for servants, with further expenses for board and wages. It also meant an increasing demand for horses, and consequently a relatively reduced possibility of feeding cattle. This was a bad circle of impoverishment and pauperization. Increasing expenses and reduced product surplus implied less saving, which led to bad farming, which meant reduced yield, etc.[79] If the ability to pay the rent was too greatly impaired, this would also mean a reduction of manorial income, but the main limit to manorial exploitation of the peasantry was the interest in the reproduction of the basis of labour duties.[80]

Large-scale manorial farming was increased not only by the effort of the old Danish nobility, but also by the new 'service nobility' introduced by monarchic absolutism after 1660. The development is reflected in the number of peasant farms, *bøndergårde,* at different points of time. While the total number is estimated at more than 75 000 in 1651, less than 60 000 were registered in 1682, and the number decreased further towards the year 1700.[81] Nearly half of these peasant farms had been crown property by the middle of the century, but this share had shrunk to 1/4 in the last decades before 1700. Now the property of new and old noblemen was double the number of peasant farms owned by the king, even if their share had been more or less equal to that of the king a few decades earlier, and in spite of the increasing significance of noble manors in addition to the ownership of peasant land. A new group of bourgeois proprietors also possessed about 1/6 of the peasant farms towards the beginning of the 18th century, and other public institutions, like parish churches, hospitals and schools, possessed about 10 %. Peasant freehold, of the restricted type existing, was quite negligible, as noted earlier.[82] Unlike the Norwegian type of tenancy, these peasant farms were run by personally unfree tenants, subject to villeinage (especially in Zealand and the islands), hard duties, compulsive residence, and the landlords' jurisdiction. The dominant noble manors constituted maximally 1 % of Danish farms, but more than 10 % of the total rent, the *hartkorn.*[53] Economically, the manors also had the advantage of not being restricted by the system of common cultivation and mixed strips to which the remaining farms of the villages belonged. Towards the second half of

[79] Cf. Olsen, *Hovedgård og bondegård,* 1957, op.cit., p. 223.

[80] Ibid., p. 224, and pp. 226–227.

[81] For 1651, see Fridericia 1889–90, op.cit., p. 514, and for the later figures see G. Olsen in *Danmarks Historie,* vol. 8, 1964, op.cit., p. 328.

[82] Cf. Olsen 1964, op.cit., pp. 330 ff.

[83] Ibid., p. 321.

the 18th century, Denmark was a country of big landed estates, where the peasantry were the dependent subjects of the landlords or the Crown.

The degree of large-scale manorial concentration was less comprehensive in Sweden than in Denmark, even if there was a parallel development.[84] Institutional home farming was stimulated by the same decisive factors as in Denmark–*säteriene,* the manorial home farms, were more totally taxfree than the scattered noble land, and they supplied duties from the peasantry–but the agrarian price depression was probably less striking in Sweden.[85] Large-scale home farming was, however, supported by the 'reduction', *reduktionen,* of 1680–1700, whereby the landed estates transferred from the Crown to the nobles during the preceding century were restored to the possession of the king. The fact is that this restoration was not based upon farm land identical to the previous transfer, but upon a purely bookkeeping calculation–the rent value (tax value) of the farms. Thus the landlords mainly delivered scattered farms, *strøgods,* to keep the manors intact, since the crown was principally interested in tax income rather than manorial farming through its own agencies.[86] The proportion of noblemen's estates was reduced from more than 2/3 to less than 1/3 of Swedish land from the middle to the end of the 17th century, but the nobility still possessed more than before the original transfer to their advantage a hundred years earlier, and now the holdings were more concentrated.[87] On the other hand, it is probable that the conditions of nature, soil and climate, being more like the situation in Norway in substantial parts of the country, made extreme concentration of scattered holdings into manors and home farms less attractive than in Denmark. The manors generally had to be smaller than the Danish ones, thereby implying less suppressive duties laid upon the shoulders of the peasantry. And scattered holdings in different parts of the country were also an economic insurance, a guarantee against local crop failure preventing effective rent collection, which could prove to be more decisive than within the more homogeneous conditions prevailing in Denmark.[88]

Thus, Sweden as well as Denmark was characterized by a dominant

[84] At the maximum 8 % of Swedish land was concentrated in large-scale noble manors–cf. Ingers I, op.cit., p. 272.

[85] Kåre Lunden (1969, op.cit., p. 232) suggests that the agrarian depression was reinforced by manorial farming. In Norway, with a different agrarian structure, the general price depression did not lead to a similar crisis among the peasantry, but probably rather stimulated the transition from tenancy to peasant proprietorship–ibid., p. 238.

[86] Cf. Heckscher 1957, op.cit., pp. 138–139.

[87] Ibid. p. 145.

[88] Cf. Ingers I, op.cit., p. 272.

rural aristocracy prior to the transformation towards 'modern' agricultural freeholds. In Sweden there was a combination of manorial farming and contractual tenancy with personally free peasants. Denmark had an even stronger concentration of large-scale estates, supplemented by leasehold based on a highly dependent peasantry. The Norwegian contrast was a system of peasant proprietorship combined with relatively free, contractual tenancy.[89]

Farm structure and land tenure

To account for the typical Norwegian farm structure prior to the transforming effects of the agricultural revolution, i.e. the enclosures, it is unnecessary to go further back than to the early 19th century. The changes that had taken place during the preceding centuries seem mainly to have been affected by demographic factors and the prevailing system of inheritance rather than by technological advances or the individualizing consequences of commercial agriculture, and did not alter the basic peculiarities. Prior to the enclosures of the late 19th century, Norwegian farms usually consisted of a relatively large number of buildings clustered together around a farm yard, a *tun*.[90] The buildings of two or several individual farms were often clustered together into a multiple abode (*klyngetun*), and the Norwegian word for a farm, *gard*, was used to indicate both a single farm and a complex of farms clustered together.[91] Especially in the west and in the northern coastal districts the multiple abode was an outstanding characteristic. Here the community fellowship generally was closer and more comprehensive than in the more fertile agricultural districts in central and eastern regions.[92] The cultivable soil was scarce, and the dwellings had to be tightly concentrated even if the farms were split up into many individual holdings. In Trøndelag and eastern Norway, on the other hand, multiple farms were often organized into single rectangular *tuns*, i.e. chains of private yards, with separate yards for the cattle of each holding.[93] In some of the central fertile valleys, the cluster of buildings could be bigger and more complicated, but still with a functional system

[89] Some of the historical causes of the peasant autonomy in Norway are analysed by Kåre Lunden, 'Om årsakene til den norske bondefridomen', *Heimen,* 1971, XV, pp. 213–222, and Lunden, *Økonomi og Samfunn,* Universitetsforlaget, Oslo 1972, pp. 54–55.

[90] The Norwegian word *tun* is etymologically identical to 'town', but means a cluster of dwellings – the 'yard' of a single farm or a multiple farm.

[91] Today the word is only used to indicate a single individual farm – as has always been the case for the same word in Swedish and Danish. The word *gard*, or *gård*, is etymologically identical to the English 'yard' (as in 'court-yard').

[92] Cf. A. Holmsen, 'Eldre norsk tunskipnad', *Heimen,* vol. XIV, No. 2, June 1969, pp. 47 f.

[93] Ibid.; cf. Bjørkvik, 'The Farm Territories', *SEHR,* vol. IV, No. 1, 1956, p. 49.

of *tuns* for each holding. In the eastern plain area, the *tun* of each holding could be completely separated from the others.[94]

The characteristic of Norwegian multiple farms was an exceedingly irregular pattern, which differed markedly from the prevalent type of rural settlement in Sweden and Denmark–the north-western European *village*.[95] The complex multiple *tun* could comprise just as many buildings, but it lacked the regularity of the village: scarcity of cultivable soil stimulated an extreme concentration of dwellings; there was no central place and no clearly marked sites with the houses symmetrically arranged on each side of a main village street. The buildings were rather scattered, and the *tun* area was divided into strips and plots in the same way as the fields and meadows, rather than into regular sites.[96]

The system of field tenure was just as irregular as the organization of the dwellings. In many parts of the country difficult climate and scarcity of soil made the two- or three-field system, with a regularly rotating 'black fallow', more impracticable than in the neighbouring countries.[97] Nor was the allotment of fields and meadows made according to the size and relative position of the dwelling sites of the village, as was usually the case in Sweden and Denmark.[98] The different parts of the cultivable fields surrounding a multiple farm, the 'infield', were distinguished according to quality and function, and were usually subdivided into narrow strips and plots, the so-called *teigblanding*, with scattered strips and plots belonging to each individual farm. The strips of each holding were generally smaller in western and northern areas, while the eastern and central (Trøndelag) farms had their territories organized into simpler and bigger plots. A second type of tenure that was notable in certain coastal areas was common ownership. This could imply two types of use–either as a rotating allocation of strips and plots, where the use of specific pieces of land rotated between the holders from year to year, or as completely communal use, where ploughing and sowing was done jointly by the holders of a multiple farm without any division into strips. This type of complete co-operation on common land even in the infield, was practised in certain

94 Ibid.

95 A 'village' is called a *by* in Swedish and a *landsby* in Danish. There is no Norwegian equivalent, since a *by* in Norwegian means a 'town', while the word *landsby* is used only about villages in other countries.

96 Cf. Bjørkvik, *SEHR*, 1956, op.cit., p. 44.

97 The widespread Norwegian system required correspondingly more manure; cf. Bjørkvik op.cit., p. 50. See further chapter 7 below.

98 Ibid., p. 49. The Swedish notions for this system were *solskifte* for the field tenure and *radby* for the usual type of village–cf. G. Berg and S. Svensson, *Svensk Bondekultur*, Bonniers, Sth. 1971, p. 56.

coastal areas close up to this century.[99] In the eastern and central regions, common ownership was usually restricted to the 'outfield', the pasture meadows, woods and peat marches. Individual ownership generally diminished the greater the distance from the central nucleus of the farm.[100] The outfield could also be divided into separate plots, but 'ownership' could nevertheless be extremely complicated: one peasant could 'own' certain types of wood, another could possess other types, a third could have the pasture rights, and yet a fourth the rights of taking fish and game. Sometimes the forest ground remained under common ownership, despite the intricate division of outfield use.[101]

A typical pattern was the prevalence of various types of tenure in concentric circles from the farm nucleus: firstly the cluster of dwellings with individual ownership of the house and its ground; secondly the mixed strips of the arable land; thirdly the mixed strips of the hayfield with common use during the cattle grazing in the autumn; fourthly common ownership of outer hayfields with rotating allocation of use; and finally common ownership and common use of woods and pastures, or common ownership with specific individual privileges and certain rights for all.[102] Even in the eastern and central districts, where individual tenure prevailed to a greater extent than the high degree of communal ownership and open-field system of the southern, western and northern areas, cultivation was often a common task.[103]

In wide parts of central and eastern Norway, especially in the valley areas, extensive farming through particular summer farms in the mountains, often at some distance from the main farm and its fields and meadows, was quite widespread. This *seter* farming comprised summer pasture where the cattle could graze from spring to autumn, and is found in many mountain regions right up to the present day.[104]

As a last expression of the extensive use of meagre natural resources, the different farms of the community also had certain rights pertaining to partial exploitation of the *almenning,* the common land belonging to the parish.

This Norwegian pattern of farm structure and field tenure was strongly based on mutual dependence and mutual adjustment. The rural communities, situated in a mountainous, wooded and sparsely populated country, were relatively isolated, with a high degree of

[99] Cf. Kristian Østberg, *Norsk Bonderett,* Vol. XI, 'Fellesskap i jord og skog', Oslo 1936, pp. 17 ff.; Bjørkvik, *SEHR* 1956, op.cit., p. 54.

[100] Cf. Holmsen, 'Norsk gardshistorie', in *Gard, Bygd, Rike,* 1966, op.cit.

[101] Cf. Østberg 1936, op.cit., pp. 57 ff.

[102] Cf. Mortensson 1904, op.cit., pp. 34–49.

[103] See Rigmor Frimannslund, 'Farm Community and Neighborhood Community', *SEHR,* vol. IV, No. 1, 1956, pp. 62–81.

[104] The *seter* farming has been on the program at the *Institute for comparative cultural research* in Oslo. Cf. Lars Reinton, *Seterbruket i Noreg, Vol. I–III, Oslo, 1955–61.*

peasant autonomy and economic self-sufficiency.[105] The pattern was very different from the regular village system of Sweden and Denmark prior to the enclosures which, in these countries, took place from the last decades of the 18th century. Clearly organized villages were the prevailing type of rural settlement in Denmark, even if single farmsteads existed in meagre areas, with scarcity of pasture and cultivable soil, especially on the moors and marsh-land of western Jutland and in certain island districts with similar natural conditions. In some Jutland counties the single farms amounted to about 1/3 of the total number of farms towards the end of the 17th century, with an average of 15 % for Jutland as a whole, while there were only about 2 % single farms in Zealand, and a fairly even distribution among the counties.[106] In the 18th century, most Danish peasants lived closely together in villages, as they had done since antiquity.[107] In some counties, mostly on the Zealand islands and in eastern Jutland, the houses of the village were placed in a circle around an inner central place, while in other districts the buildings were arranged symmetrically on both sides of a long main street, or, more unusually, in one single row along a river, a beach or a bog.[108] The village land was divided into a lot of relatively homogeneous parts, and each farm was given a narrow strip of each part, corresponding to the position of its dwellings in the village. The fields could also be arranged in radiating plots with the nucleated village at the centre, especially if the village was dominated by the manor of a landlord. Even if each strip was cultivated by the leaseholder himself, arable, crop rotation, field work and utilization of pasture had to be closely coordinated. This coordination was ensured by detailed village regulations comprising farm work, mutual obligations and internal jurisdiction. Under the influence of the landed aristocracy, the peasant holdings within the village were made quite equal in size, with similar rent and duties, by the organization of field tenure.[109] During the 17th and 18th centuries, these originally quite self-governed village units were more and more strongly dominated by the big estate owners, the lords of the land. In the second half of the 18th century, the fellowship of the village, as it was expressed by the formal regulations of communal work and life, was incorporated into the system of control exercised by the dominant manors and landed estates. The regulations

105 *Cf. Holmsen, 'The Old Norwegian Peasant Community', SEHR,* IV, No. 1, 1956, pp. 17–32.

106 Cf. Svend Aakjær, 'Bosættelse og Bebyggelsesformer i Danmark i eldre tid', in *Bidrag til Bondesamfundets Historie,* II, Oslo 1933, tables on pp. 176–179.

107 Ibid., pp. 109 ff.

108 Cf. Fridlev Skrubbeltrang, *Den Danske Bonde 1788–1938.* Copenhagen 1938, p. 12

109 Cf. Paul Meyer, *Danske Bylag,* Copenhagen 1949, p. 296. Meyer discusses the development of mixed field tenure from the *årsskifte* (pp. 259–61) through the *bolskifte* (pp. 261–64) to the most consistent solution, the *solskifte* (pp. 264–65).

included the privilege of corporal punishment at the hands of the landlords and their bailiffs.[110]

The Swedish system of rural settlement and land tenure seems closer to the Danish pattern than to the Norwegian one, even if Swedish villages generally were smaller, with fewer dwellings.[111] The parish was also more important than the village as an organizational unit for local government, and from the 17th century onwards it took over a lot of secular functions.[112] There was also a greater frequency of single farmsteads in Sweden than in Denmark, although most peasants were settled in nucleated villages all over the country. The distribution of mixed strips and plots was quite strictly regulated, and so was the organization of field labour, crop raising and pastures. Prior to the 18th century, old village regulations had been quite detailed for each individual village, even if they were confirmed at the regular parish meeting. They were rather autonomously constructed, and differed a lot according to local conditions.[113] During the Middle Ages, the land tenure of the villages became regulated after the Danish pattern, so that the distribution of mixed strips among the peasant holdings corresponded to the position of each settlement within the village.[114] In some areas the utilization of certain plots rotated among the holders right up to the 18th century, while the forest often was common land, even if the shares of the outfield amenities corresponded to the size of the village sites.[115] In the 18th century new compulsive village regulations were introduced and harmonized to stimulate a more 'rational' agricultural household. A model of village rules was issued by the State in 1742, with a public inspector in each village to lead enclosures, coordinate farm management, and oversee land drainage and fencing.[116] Thus the regulation of agriculture and village order in Sweden and Finland was conducted under the leadership of the Crown, and not left at the local landlords' discretion as in Denmark. The Swedish village regulations, however, aimed at winding up the strip farming, and should partly be seen as an alternative to the parallel enclosures, which were officially introduced in Denmark towards the end of the century as well.[117] Public modernization extended to the agricultural sector, in connection with the intrusion of the market into rural areas, belongs to the following chapters.

[110] Cf. Skrubbeltrang 1938, op.cit., pp. 15 f; *Danmarks Historie,* op.cit.

[111] Cf. Sigurd Erixon, 'Hur Norge och Sverige mötas', in *Bidrag til Bondesamfundets Historie,* II, Oslo 1933, pp. 183–299, esp. p. 217.

[112] Cf. Meyer 1949, op.cit., p. 56, p. 63, and p. 66.

[113] Cf. Jutikkala 1963, op.cit., pp. 263–264; Ingers I., op.cit., pp. 46–53, esp. p. 52.

[114] Cf. Ingers' discussion of the *solskifte,* Ingers I., op.cit., pp. 48–49.

[115] Ibid., pp. 49–50.

[116] Meyer 1949, op.cit., pp. 75 ff.; Jutikkala 1963, op.cit., p. 265.

[117] Cf. Jutikkala, op.cit., pp. 262 ff.

The original contrast between Norway on the one hand and Sweden and Denmark on the other, as to systems of rural settlement, is really striking. The Norwegian peasantry have never seen any parallel to the detailed village regulations that prevailed in the other Scandinavian countries. This fact may be explained by conditions of land ownership, with little landlord control, and with the centre of political authority located quite far away, in another country, since the Middle Ages.[118]

Social order and land disposal

We shall now compare, rather schematically, four aspects of the social and institutional order in so far as they concern the system of access to land prevailing in the Scandinavian countries, namely social hierarchy, relations of authority, economic obligations, and legal institutions. Together with the land tenure systems, these factors build an interrelated agrarian formation, and they will be sketched for each national setting.

Generally, the *ancien régime* of each Scandinavian country was an estate society with a sharply delineated system of status. Social estates differ from social classes, as they are defined not only by socio-economic criteria, but by legal and political ones as well. The gradations between the few superior and the many inferior were authoritatively established. Even if there were social classes within the estate society, only the upper class was capable of concerted action over the whole area of society, while the peasantry was either too localistic or too split by segmentation and patronage.[119] The formal character of the estate society was particularly strong in Sweden, where the Assembly of the Estates was the centre of political organization. The principle was weaker under the absolute monarchy of Denmark–Norway, but even there the population was divided into clearly differentiated groups with relatively fixed rights and duties of a juridical character. The social delineation was thus more than mere social distance based upon conventions or economic differences. The King was the supreme power, and the privileged nobility formed the recognized summit of the estate society. Just below the summit followed the higher clergy and civil officialdom, and then the bourgeois proprietors, who might be commercially privileged through mercantilistic legislation. The peasantry were in a different and inferior world, even if there was important class differentiation within the peasantry as well. The rural landless were

[118] The natural conditions of Norway, with less regularity in crop rotation, also carried weight in this respect–cf. Holmsen, SEHR 1956, op.cit.

[119] In this way Peter Laslett talks about the English old regime as 'A One-Class Society', cf. Laslett, *The World We Have Lost,* 2*e*d., Methuen, London, 1971, Ch. 1, pp. 23–54.

close to the bottom, as were the servants and apprentices of the urban world. The bottom was the 'poor'–the paupers, the beggars, the lepers.

In *Norway* the general hallmarks of an estate society, in political and legal terms, were much weaker than in the other Scandinavian countries, especially compared to Sweden. The summit of the estate society, the apex of the social pyramid of the *ancien régime*, was virtually non-existent towards the end of the 18th century: there was no strong landed nobility with administrative power and political privileges.[120] True, noble titles were not formally abolished until 1821, but most of the aristocratic families had no basis in the possession of land, which was owned mainly by bourgeois proprietors, peasant freeholders, and civil servants. Around 1800 only a small proportion of crown estates was left, and there existed only three noble estates with certain privileges–one barony in a western county, and two counts' estates *(grevskap)* south-west of Oslo. These were established after the introduction of absolute monarchy in Denmark–Norway, but were just exceptions to the general insignificance of the Norwegian nobility.[121] After 1661 the provincial officialdom was formally separated from the old nobility and connected to the rising bourgeoisie. Through bourgeois civil service, the system of revenue farming was replaced by fixed salaries with landed income. The top layer in Norwegian society, the economic and administrative elite, was formed by bourgeois proprietors and merchants, mill- and mine- and factory owners, and non-noble clergy and civil officials. This dominant bourgeoisie of commercial magnates and bureaucrats was socially segregated from the lower classes, often tied together by family relationships, and closed against the lower strata. The alien character of the upper class was culturally emphasized by the fact that their ancestors had mostly immigrated from abroad, especially from Germany and Denmark, during the 17th century. Norway thus lacked, contrary to Sweden and Denmark, a native upper class to absorb the alien elite. The army officer corps was even formally closed to sons of peasants.[122] There was then an extremely wide social distance to the lower classes which underlined the social coherence of the elite. In some matters this distance was sanctioned by official decrees, but the attitude of all groups was regulated by fixed conventions of social intercourse, whether these had a legal character or not.[123]

In quantitative terms, less than 50 000 of the Norwegian population belonged to the 'bourgeoisie' in 1769–civil servants or the bourgeois

[120] Cf. Ingrid Semmingsen, 'The dissolution of estate society in Norway', *SEHR,* vl. II, No. 2, 1954, p. 168.

[121] Axel Coldevin, 'Enevoldstiden', *Vårt Folks Historie,* vol. V, pp. 36–39.

[122] This was amplified in a royal decree of 1772–cf. Steen, *NFLH,* vol. VI, pp. 302–303.

[123] Cf. Semmingsen, *SEHR,* 1954, op.cit., p. 170.

commercial class with their wives and children–while more than 650000 were 'non-bourgeois'–peasants, sailors, servants, and the 'poor'. Civil officialdom constituted between 1/3 and 1/4 of the bourgeoisie.[124] Still in 1801 more than 4/5 of the population belonged to the 'peasant society', and about 9/10 of the population lived in the countryside. The independent farmers–freeholders and tenants with their families–comprised about 40 % of the population, and the rural landless–crofters, servants and day labourers–slightly more, about 42 %.[125]

There was a clear differentiation within the 'peasant society' too, even if the social structure of the relatively isolated peasant communities varied greatly according to local ecological conditions, institutional arrangements, and land/people ratios. Within a local rural setting the 'yeomen class', the bigger tenants and the allodial peasant proprietors (*odelsbønder*), constituted a relatively independent upper class, followed by the more modest tenant farmers, the crofters, the servants and day labourers, and the paupers–down the social hierarchy–according to the degree of agricultural responsibility and control of land. The crofters, or cottagers (*husmenn*), became a vast rural lower class from the middle of the 18th century, following the population growth and the transition towards peasant proprietorship.[126] The population surplus of pre-industrial society had to be absorbed by agriculture, and crofters were hired by the freeholding peasantry as a labour force in land clearing, forestry and farming. Some crofters got a piece of land in return, a small plot with a house to work and live on. The crofter had labour duties on the main farm in return for a fixed salary, and he paid a leasehold fine for his piece of land, but he kept the produce, which was inadequate to sustain himself and his family. He could also increase his crop by clearing new arable land around his plot, but his position was insecure, since the master owned the land. The landholding crofter was thus neither a freeholder nor a tenant, but something between a servant and a leaseholder.

From the second half of the 17th to the end of the 18th century, the proportion of crofters increased from approximately 1/7 to 1/3 of the whole peasant society. This growth has been connected to the increase in lumbering and sawmill operations during the 18th century, with a growing need of a stable potential labour force. The specific solution

[124] This is an approximate estimate from the census of 1769–cf. Steen, *NFLH*, vl. VI, p. 300.

[125] Cf. Steen, *Det gamle samfunn*, 1957, op.cit., p. 56. Steen's reservation, in a critical evaluation of the census data from 1801, is that some persons with two or more occupations might have been counted more than once. But the rough proportion of overall categories is not fundamentally distorted by this.

[126] Cf. Steen, *NFLH*, VI, pp. 197 ff.

of the problem was also stimulated by the public tax policy from 1752, implying that the assessed farm rent was not to be increased even if new plots were cleared and cultivated on registered land.[127] From now, and during the whole 19th century, the crofters formed a solid rural proletariate as lifetime land labourers and farmland clearers for the peasant proprietors (cf. chapter 6 below).

The extreme regional contrasts of Norway affect not only the system of land tenure, but the rural social structure as well. During the 17th and 18th centuries the increasing population had to be absorbed by the rural communities. This changed their structure, since there were no big urban centres, and no industries, to absorb the surplus population.[128] Along the western coast, in the north, and in the mountain regions, outside work like fishing, seafaring, and efficient utilization of the mountain pastures made it possible to subsist on smaller holdings. This allowed an extreme subdivision of farms to take place by the inheritance system, and it also stimulated more social equality in the farm community.[129] Crofters, labourers and servants were often younger sons and daughters of the peasants, and lived together, nearly like family-members, as close relatives of the main farmers. These small, tightly packed peasant communities showed a relative equality in socio-economic status. In the wealthier agricultural parishes in the centre (Trøndelag) and the south-east, there was a greater social distance. The subdivision of farms came to an end much earlier. It was possible to acquire a crofter's allotment on the parental farm for only a few generations. Younger sons and daughters of the peasant farmers migrated to other areas, outside the home parish or within it, rather than become crofters and servants in their father's or elder brother's household. Thus the basic solidarity in these districts seems to be characterized by a dependence-responsibility relation rather than a kinship-equality base.[130] The 18th century timber commerce of the eastern and central regions also stimulated an increasing number of labourers and crofters on the outskirts of the main farms, in a non-equal, patriarchal structure.[131]

The Norwegian 18th-century peasantry were not subject to personal dependence of a semi-feudal kind. As for the general *relations of*

[127] Ibid., pp. 200–201; cf. Koht, *Norsk bondereising*, 1926, op.cit., p. 208.

[128] The population growth from about 1500 to 1665 had been equally strong, but during that period the population surplus either cleared new farms or settled on deserted farms, *ødegårder,* which had been abandoned after the Black Death–cf. Holmsen, *SEHR,* 1956, op.cit., p. 22, note 4; on the population growth in the 18th century, see M. Drake, 1969, op.cit.

[129] Cf. Rigmor Frimannslund, *SEHR,* 1956, op.cit., p. 64 and p. 69; and Holmsen, op.cit. p. 22, note 4.

[130] Ibid.

[131] Ibid.; Koht 1926, op.cit.

authority, it has often been stressed that Norwegian tenants were bound to their landlords by a free, contractual relationship, and never by a state of serfdom.[132] The peasants were not subject to the landlords' jurisdiction, and duties were regulated to the advantage of the tenants in the latter part of the 17th century, with guaranteed leasehold for life and a right of inheritance of the lease.[133] The only exceptions to these regulations were the peasants under the barony and the two counts' estates, who were more dependent. The peasantry also had a relatively high degree of managerial responsibility, since the proportion of large-scale farms worked by farm labourers under a bailiff was quite small. There were even official sanctions, although not totally efficient, against transforming tenant farms into unified manors.[134] The Crown had obvious reasons for the preservation of a smallholding peasantry at the expense of a strong landed upper class in Norway.[135]

The managerial conditions of farming were not particularly different whether the peasants were tenants or freeholders. Basically, they were connected to the non-peasant world by formal and legal relations of authority rather than by structures of patronage. Paternalistic integration was stronger within the peasant society as such, particularly in the areas with a marked polarization of wealthier freeholders and dependent landless, while the corporate links of kinship integrated the communities in the south and west.

In the 18th century there was a scarcity of servants and labourers in the rural areas, and by a public decree of 1754 duty to serve was made compulsory for all landless people of peasant origin.[136] Thus farmers and mine- and sawmill owners were guaranteed cheap labour under the control of the county bailiffs, by a kind of compulsive residence system, which was not finally abandoned until 1854.[137] The crofters also worked and lived at the landowning farmers' discretion, often without a written contract regulating mutual rights and duties. An official decree in the middle of the 18th century intended to make the crofters into tenants with leasehold for life and a fixed contract, but this was withdrawn by the government after two years' opposition from the proprietors and landowning peasantry.[138] The asymmetrical

132 Cf. Oscar Albert Johnsen, *Norges Bønder,* Aschehoug, Kristiania 1919, p. 40; Bjørkvik 1958, op.cit., p. 90.

133 Regulated by law in 1684–85 – cf. Steen, *NFLH,* vol. V, pp. 262 ff.

134 Cf. Steen, op.cit.

135 This aspect will be further analysed in the next chapter, in the discussion of the transition towards peasant proprietorship.

136 Cf. Steen, *NFLH* VI, p. 303; Johnsen 1919, op.cit., p. 275. The provision also guranteed available soldiers to the monarchical power.

137 Johnsen, op.vit., p. 275. It should be noted that the dependent landless ordinarily could change service every fourth year.

138 Cf. Steen, *NFLH,* VI, pp. 200–201; Johnsen, op.cit., pp. 253–254.

power relations dominating the internal structure of the peasant society thus seem to be of another kind, characterized by a more personal type of dependence, than the formal relations between peasant and non-peasant groups. This picture is, of course, extremely incomplete without a discussion of the *economic obligations* connecting the peasantry to the wider societal order.[139]

Except for the taxfree noblemen's estates, all Norwegian farms–whether they were used by tenants or peasant freeholders–had certain *public duties*. The farm was always a part of greater administrative and political units, the *skiprede* (an old defence unit for the mobilization of ships), the parish, the district (*herred*), the county (*amt, fylke*), the state. This political belongingness meant obligations and duties of an economic character.[140] It was the *farms* which had to supply the state and the church with duties and services–like the public taxes and the tithe to the church.[141] Compulsory military service was connected to the *farm* from the Saga Age to the end of the 18th century. Social contributions of a public character were also connected to landed property and adjusted to the size of the farm. From the Middle Ages there were two, and later three, tax categories–for the 'full farm' (*fullgard*), the 'half farm' (*halfgard*), and the 'quarter farm' or 'deserted farm' (*fjordingsgard/øydegard*).[142] From the middle of the 17th century this system was replaced by a more flexible system whereby the tax was partly adapted to the assessed land rent (*landskyld*) of the farm.[143] In 1661 a big land commission revised the land rent value as the basis of assessment, in a fixed relation to the yields of the farm.[144] The *landskyld* was originally a leasehold duty, a rent to be paid by the tenants, but it was soon attached to *all* farms as a general measure of farm value and a basis for both public taxes and the administration of estate inheritance.[145] The 'allodial tax' and 'tenant tax' (*odelsskatt* and

139 The approach to this aspect will be of an institutional rather than a quantitative character.

140 Cf. Sverre Steen, 'Bisysler for bøndene–arbeid og markedsferder', in *Bidrag til Bondesamfundets Historie,* Vol. I, Aschehoug, Oslo 1933.

141 The tithe, a tenth of all peasant produce, formed the old economic base of the church. Generally, as concerns the duties towards the state, there was a gradual change from labour services to taxes. The tax and rent changed from basic foodstuff to general merchandise in the 15th and 16th centuries, and gradually into money in the 18th and early 19th centuries–cf. Sigvald Hasund, *Bønder og stat under naturalsystemet,* Olaf Norlis Forlag, Kristiania 1924, esp. p. 34 and p. 91.

142 Ibid., pp. 46 ff.

143 By the royal tax decree of 1647–cf. Bjørkvik 1958, op.cit., p. 30.

144 The norms of evaluation were based upon crops, livestock and 'accompanying amenities', which were measures of the use value like the later 'registered rent', the *matrikkelskyld.*

145 Cf. Bjørkvik 1958, op.cit., p. 27. Cf. note 49 above. See also *Kulturhistorisk Leksikon for nordisk middelalder,* vol. X, Oslo 1965, pp. 277 ff.

leilendingsskatt) became annual extra taxes in the 17th century, but even the *leilendingsskatt* soon became a regular tax on land utilization also for freeholders.[146] Until the middle of the 18th century, most of the taxes and rents were delivered in agricultural products, but at that time there was a gradual change to money rents, mediated by increasing deliveries to the town markets. One exception was the tithe to the priest, which was still paid in kind. Some public duties were paid as obligatory labour towards the middle of the 19th century.[147]

The peasants also had a duty of conveying travellers–especially the landlord–including the king and his servants, and to maintain them when they passed through the districts. The conveyance of noblemen and public officials was free of charge, while the peasants were paid for the conveyance of private travellers. This system lasted, roughly, to the second half of the 19th century.[148]

A last type of public charge on the peasant economy was the collection of fees for official activities. This was an important source of supplementary income to the meagre salary of provincial officials, and also, as we shall see, a source of endless quarrels and complaints from the peasantry to the king before and during the period of absolute monarchy.[149] The *private obligations* of the leaseholding peasantry were of another kind, even if they too, more or less effectively, were regulated by public law. These obligations were based upon the control of land as a factor of production, and the most important in Norway were the leasehold or tenancy rents. The new tenant first paid an entry fee, the *førstebygsel* ('first lease'), and then a regular lease fee every third year, the *tredjeårstake,* as a reminder of the earlier system when the leasehold contract had to be renewed every third year. In addition, the tenants paid the annual rent, the *landskyld*, to the landlord.[150] This ground rent generally was higher the stronger the demand for agricultural land.[151]

Some of the tenant peasantry also had labour duties on noblemen's estates and on the farms of public officials, but in most places the obligations towards the officials no longer applied, or were replaced by a monetary fee, in the latter part of the 18th century.[152]

146 Ibid., pp. 29–31. The classifications of the tax register therefore give a very invalid expression of the proportion of freehold.

147 Especially the obligatory maintenance of roads and bridges–cf. Steen, 'Bisysler . . .', 1933, op.cit.

148 Ibid.

149 The classical and most coherent account is still Koht, *Norsk bondereising,* 1926, op.cit.

150 Cf. Holmsen 1966, op.cit., pp. 126 ff. Cf. also the previous discussion of ownership relations and land tenure.

151 Cf. Bjørkvik 1958, op.cit., p. 14.

152 Cf. Steen, "Bisysler . . .", 1933, op.cit.

The degree of direct access to vertical markets is a second major type of economic relationship defining the social situation of Norwegian peasantry. In the 18th century some produce was sold on the market-place mainly because taxes and rents were increasingly claimed in money, although a substantial part of the rural trade still prevailed as direct commodity exchange. However, important privileges of commercial market trade in basic products were given to towns and urban merchants, and the guild organizations had exclusive rights to the exercise of crafts. These mercantilistic legal provisions had their roots back in the Middle Ages, and were the most typical remnants of a Norwegian 'estate society'.[153] But they were also a legal expression of the new capitalistic development. The commercial bourgeoisie was given the exclusive right of sawmill operations for export, by special 'quantity sawmills' (*kvantumsager*) as early as 1688.[154] Commercial privileges prevailed into the second half of the 19th century. Thus the economic adaptation of Norwegian peasantry met with a specific institutional framework–the consequences of which will be traced in later chapters.

Another aspect of the *institutional order* seems rather to belong to the self-defence of the peasant society. This is the old Norwegian *odelsrett*, the 'allodial privilege' of the peasant farmers, which prevailed, although in a slightly changing or fluctuating manner, throughout the centuries. The peculiar Norwegian *odelsrett* (which still exists) implies that a family that has possessed a farm for a certain number of years acquires a kind of prescriptive right, through time-honoured custom, to the farm. If the farm is sold to someone else, the members of the family, in a fixed order, then have a right to buy the farm back within a certain number of years. Both the period required for the acquisition of the privilege and the time-limit for re-acquisition were often changed by public law, but the attacks on the institution as such were always rejected by the authorities.[155] The institution of *odelsrett* means that the disposition of land is not primarily regulated by commercial market rules, but by rules for distribution within family groups. The proportion of land without allodial restrictions, and the rigour of the allodial principle itself, is thus a measure of the degree to which mercantile domain

153 Cf. Semmingsen, *SEHR* 1954, op.cit., p. 169.

154 These were established by a royal decree of 6 September 1688, originally on the ground that there was a risk of deforesting.

155 The *odelsrett* was attacked at the end of the 18th century on the ground that it obstructed freedom of trade in land–cf. Knut Robberstad, *Fyrelesningar um Rettssoga i millomalder og ny tid*, vol. II, 4 ed., Universitetsforlaget, Oslo 1966, pp. 126 ff., and Robberstad, 'Odelsrett', in *Kulturhistorisk Leksikon for nordisk middelalder*, vol. XII (1967), pp. 494 f.

It should be noted that the holders of crofter plots were not entitled to this right.

and capitalistic principles dominate the agrarian system.[156] The Norwegian *odelsrett* was supplemented by the *åsetesrett*, which was the right of the eldest son to take over the entire farm undivided and at a modest price. This principle was a barrier against extreme subdivision of farms and impoverishment of the farmers, and proved to be particularly effective in eastern and central parts of the country. Here it also produced a rural proletariate from disinherited brothers and sisters because of population growth and scarcity of land. But it fortified the social and economic backbone of the freeholding peasantry.[157]

The landed society in *Denmark* was strikingly different from that of its subordinate twin country in the north, even if it existed within the same overall political framework, and even if the basic hierarchical status groups were more or less the same.[158] Since the old nobility had dominated the rural scene to a much greater extent than in Norway, the change induced in the age of absolute monarchy was more dramatic. The Danish landed nobility had been politically shielded by legal privileges and a sharply delineated system of status, which also meant protection against the pressure from other groups. This favourable position was broken down by the introduction of absolute monarchy in 1660. The old nobility now met competition and pressure through the advance of the bourgeoisie and of the new civil officialdom of absolutism. It has also been argued that the position of the landed nobility was weakened by economic factors.[159] The nobles withdrew from trade and other capitalist activities to concentrate on the management of their landed estates, and left the commercial world to the strongly competitive foreign capital and enterprise. This one-sided agricultural base was more sensitive to the fluctuations of the market, and the agrarian depression from the 17th century decisively enfeebled the old aristocracy.[160] Thus the Danish landed nobility, which had been at its peak as a political elite between the Reformation and the absolute monarchy, was already reduced to a rather insignificant group by the first decades of the 18th century.[161] This decline of the old nobility did not mean that aristocratic influence disappeared altogether. The

156 Because of the complicated system of shared rent and partial ownership of registered farms, it is extremely difficult to give a valid estimate of the real proportion of allodial land in this early period.

157 Cf. the judgement passed on the origin of the institution in Holmsen 1966, op.cit.

158 Cf. the rough distribution of the main groups in the population statistics at the beginning of the present chapter.

159 Cf. Svend Aage Hansen, *Adelsvældens grundlag*, Gads forlag, Copenhagen 1964, pp. 273 ff. A review, with some evaluation of the use of sources, is given by K. Svastoga in *Nationaløkonomisk Tidsskrift*, 103, 1965.

160 See Lunden 1969, op.cit.

161 According to S. Aa. Hansen, this was brought about by the combined effect of the agrarian depression in the economic field and the absolute monarchy in politics– cf. Hansen 1964, op.cit., p. 276.

first kings of absolutist reign established a new quasi-feudal nobility both to lend lustre to the monarchy and to split the old aristocracy.[162] The institution of privileged counts and barons, from 1671 onwards, did not really imply that crown estates were made into fiefs used by noblemen. It was rather private property which was raised to the honour of 'fiefs'. Rich landowners, often of the former aristocracy, were given tax privileges and partial jurisdiction over their peasants and servants, while, in return, their landed property should revert to the Crown if the privileged family died out.[163] Contrary to the old nobility, this new aristocracy was thus established to support the Crown.[164] Unlike the continental system, the new fiefs (*len*) were neither independent nor hereditary, but rather civilian-military offices connected to the monarchical power.[165] The obligation of service and fidelity to the Crown, and the agreement that the land should revert when the family become extinct, was exchanged against tax exemptions and local jurisdiction over the dependent peasantry. It is also likely that the establishment of count's estates and baronies accentuated the general tendency towards unified manor farming in late 17th and 18th century Denmark.[166] If the gap, in power and social status, between the peasantry and the lords of the land was disproportionately greater in Denmark than in Norway, the polarized differentiation within the peasantry itself seems correspondingly smaller. A substantial proportion of the landless cottagers and servants in Denmark was directly subject to the landed upper class, like the landholding peasantry, and did not form a distinct rural lower class tied to peasant proprietorship.[167]

As concerns the *relations of authority* in the Danish agrarian struc-

[162] Cf. Louis Bobé, Gustav Graae and Fr. Jürgensen West, *Danske Len*, Copenhagen 1916.

[163] Ibid., p. XXI.

[164] There has also been an old debate about the peculiarity of the earlier danish feudalism–cf. Kr. Erslev, 'Europæisk Feudalisme og dansk Lensvæsen', (Dansk) *Historisk Tidsskrift*, R. 7, B.2, Copenhagen 1899–1900, pp. 247–304.

[165] Cf. Bobé, Graae and West 1916, op.cit., p. XIX.

[166] This is suggested by Bobé et al. 1916, op.cit., p. XXV ff; cf. Olsen, *Hovedgård og Bondegård*, 1957, op.cit.

[167] We have to make some reservations here. The crofters, *husmenn*, were often subject to labour service to the manor and lived within communal land of the village, while the lodgers, *inderster*, could be labourers in service of the leaseholding peasant, living in a house placed on the land of the individual farm–cf. Danstrup & Koch (eds.), *Danmarks Historie*, vol. 9, 1721–1784, by Svend Cedergraen Bech, 1965, p. 260. There was a growing number of crofters in the 18th century, owing to population growth, closing down of farmsteads, and leaseholding peasants who could not fulfil their obligations. Around the year 1800 the number of crofters and lodgers was double that of the peasant farmers–cf. ibid., pp. 260–261.

ture, the peasantry was far more dependent–as personal subjects at the landlords' discretion–than their Norwegian counterparts. Although their relation to the landed upper class was of a less free contractual character, their tenancy contracts included a right to leasehold for life. But a great proportion of the peasants had to abandon the lease because they were unable to fulfil the contracts. It is generally estimated that about 1/4 of the tenants forfeited their leasehold and ended, deeply in debt, as landless labourers *(daglejere* and *inderster*), while an additional third left their farms because of sickness or old age. Thus only about 40 % of the tenants kept their land until they died.[168] The managerial responsibility of the leaseholding peasantry was also extremely restricted–partly because of the system of mixed strips and common tenure, and partly because of the strict village regulations which were controlled and implemented by the landlord and his bailiffs.

The personal dependence of the peasantry was codified by specific institutional arrangements tying the peasants to the landlord's estate. The *vornedskab* implied that the peasants could not leave the estate on which they were born without the landlord's permission, and that they could be brought back by force if necessary. The *vornedskab* originated in the 15th century, and was concentrated to Zealand and the southern islands, to secure the necessary labour force. The institution was formally abolished in 1702, since an increasing population made the economic need of serfdom less urgent. The abolishment was also intended to stimulate immigration to Zealand from the less unfree provinces, and did not include peasants who were born on Zealand before the inauguration of King Frederik IV in 1699. Thus the policy of abolishment had quite distant and long-range effects. There are also instances of new *vorned*-agreements between peasants and landlords after the formal abolition.[169]

A new compulsory residence system, the *stavnsbånd*, was introduced in 1733, and made more rigorous later.[170] It was intended to secure the necessary labour force as well, and implied that all males liable to military service were forbidden to leave their landlord's estate. When the *stavnsbånd* was introduced it thus covered males between 18 and 36 years of age, but only two years later it was extend to those between 13 and 36, and in 1742 further to those between 8 and 39 years of age.[171] The compulsory residence system was introduced against the

[168] This is estimated from the leasehold protocols–cf. Skrubbeltrang 1938, op.cit., p. 17.

[169] Cf. G. Olsen, 'Det unge Enevælde 1660–1721', *Danmarks Historie,* vol. 8, Copenhagen 1964, pp. 391–395; cf. also Skrubbeltrang, *SEHR* 1961, op.cit.

[170] The system comprised the entire kingdom of Denmark, except Northern Slesvig, the Island of Bornholm, and some few coastal provinces.

[171] *Danmarks Historie,* vol. 9, '1721–1784', op.cit., pp. 189–192.

background of an agrarian depression, at a time when the landed interests dominated the cabinet.[172] It was a kind of compromise between the landed upper class and the monarchy: an obligation of the landlords to supply the army with soldiers, against a right to keep any conscript peasant in compulsory labour service on the estate. The landlord also had a practically unrestricted right to punish and castigate, provided the peasant was not directly crippled or killed (which alas happened too).[173] The landlord further had local jurisdiction (*birkerett)* over his peasants, and he disposed of the right of conscription. These amenities accompanied the control of land. The threat of conscription in the hands of the landlords could be used to force the peasants to take over the tenant farms, while the *stavnsbånd* was of course the main method to secure this.[174]

Some fifty years later the liberal ideology of expanding capitalism came to dominate Danish agrarian policy, and there was a period of great reforms from the middle of the 1780s. Against the background of bad times, the unfree position of the peasantry was defined as an obstacle to economic development and enterprise: the peasantry were emancipated from above.[175] The weak peasantry were legally protected by a decree regulating the tenancy conditions in 1787, and the compulsive residence system was abolished by a decree in 1788.[176] By a decree of 6 December 1799 the last remnant of villeinage, the *hoveri* –compulsive labour duties on the landlord's manors–was restricted, regulated, and partly replaced by a money rent.[177] But several decrees in the early 1790s to some extent permitted the transformation of peasant farm-land into crofter plots, and the crofters were excluded from the advantages gained by the leaseholding peasantry.[178] Thus the peasant liberation, together with the parallel enclosures, also stimulated a social differentiation within the peasant society itself.

Most of the economic obligations of the peasants were defined by

[172] See Hans Jensen, *Dansk Jordpolitik 1757–1919,* vol. I, Nordisk Forlag, Copenhagen 1936.

[173] Cf. Gunnar Olsen, *Træhesten, hundehullet og den spanske kappe,* Copenhagen 1960.

[174] About 3/4 of Danish land belonged to the landed estates prior to the great agrarian reforms. 13 % of the land of the estates were farmed directly under the manors, while the remaining part of the estates was 'peasants farms' which the landlords was obliged to let on lease–cf. Jens Vibaek, 'Reform og Fallit, 1784–1830', *Danmarks Historie,* op.cit., vol. 10, Copenhagen 1964, pp. 57–59.

[175] Cf. hans Jensen 1936, op.cit., and Edward Holm, *Kampen om Landboreformerne i Danmark i Slutningen af 18. Aarhundrede,* Copenhagen 1888.

[176] The decrees were issued on 18 June 1787, and 20 June 1788–cf. Robberstad, *Rettssoga,* II, op.cit., p. 126 ff. See also Hans Jensen 1936, op.cit., p. 143 and p. 153 respectively. It is important to note that the peasant liberation occurred parallel to the enclosure movement, which will be discussed in the next chapter.

[177] Hans Jensen, op.cit., pp. 201–202.

[178] Ibid., pp. 206 ff., and p. 217.

their subordinacy to the lords of the land.[179] The leaseholders paid an entry fee (*indfaestning*), which was usually quite modest. Then they paid a land rent, the *landgilde,* which was an annual payment in products of the soil–usually about 20–25 % of the yield, or 5–6 times the seed corn, which was rather heavy. In addition they paid the tithe (a tenth of the produce) to the church; but the priest often got only a third of this, while the remaining part was appropriated by the landlords.[180]

The heaviest obligation, however, was the obligatory labour duties, the *hoveri-plikt,* on the manorial estates of the landlords. About 200 days annually were not uncommon, and concentrated to the spring work and harvest season, when the damage to the peasant's own farming was greatest. Some peasants sent their own servants, but then had to pay their expenses from meagre plots. Peasants who were lucky enough to live too far away from the manor could replace their labour duties by a money rent. Thus about 2/3 of the Danish peasants had compulsive labour duties by the middle of the 18th century, while the peasant freeholders, few as they were, had a fourth of the ordinary obligation.

In this situation, the formal access to the vertical market was rather irrelevant, since the agricultural surplus was extracted and controlled by the landed proprietors.[181]

Contrary to the absolutist form of government in Denmark–Norway, 18th-century *Sweden* was politically organized into estates. The monarchy was balanced against the Assembly of Estates–the nobility, the clergy, the bourgeoisie, and the peasantry–between which there were clearly defined political boundaries. There was also a wide social gap between the main status groups, even if there was a tendency towards greater mobility, and an increasingly broader base for the recruitment of the elite, from the end of the century.[182] The dominance of the landholding nobility had been limited by the Reduction of noble estates

[179] Cf. Skrubbeltrang 1938, op.cit., chapter I; Vibaek, Vol. 10 of *Danmarks Historie,* op.cit., pp. 58–59.

[180] Cf. ibid., p. 24. (Skrubbeltrang).

[181] Even when the peasants produced an additional surplus to be sold at the market, as they often did on a modest and local scale, they were met by joint agreements between the bourgeois commercials to press the prices down–cf. Andreas Jørgensen, 'Borger og bonde på Møn i 18. århundrede', *Arkiv,* vol. I (1966). It should also be stressed that a growing part of the *hartkorn* became bourgeois property–increasing from 23 % to 31 % in the period 1730–1766–cf. *Damnarks Historie,* op.cit., vol. 9, p. 270. Subsidiary trade and craft, outside agriculture proper, was generally less important then in Norway.

[182] The practicable standard career, over three generations, was from peasant to priest to civil servant–cf. Sten Carsson, *Svensk ståndscirkulation 1680–1950,* Uppsala 1950. See generally Sten Carlsson, *Ståndssamhälle och ståndspersoner 1700–1865. Studier rörande det svenska såtndssamhällets upplösning,* Gleerup, Lund 1949.

from 1680, but the absolute monarchy. 'the monarchy by the grace of God', lasted only until the death of King Charles XII in 1718. From now on the monarchic power was reduced for the benefit of the committees of the Assembly of Estates, especially the Estate of the Nobility. The aristocracy was politically dominant, in spite of the theoretical balance between the four estates.[183] From 1723 the noblemen got a formal monopoly of the highest appointments in the service of the state, and their agrarian privileges were practically unshaken–the most important of which was the exemption from those basic taxes that formed the ordinary revenue of the crown. The unified manors, inhabited and farmed by the owners themselves, were totally exempt, while other noble lands as a rule paid at half-rates.[184] During the 18th century nobles also bought non-noble land, especially in the fertile southern provinces, where large-scale manorial farming was stimulated by market production. On the other hand, the clear distinction between noble and non-noble land was obliterated, because noble land was bought up by bourgeois proprietors throughout the century.[185] The privileged position of the Estate of the nobility was challenged by anti-aristocratic politics – from the lower clergy, the lower urban bourgeoisie, and the peasantry, around the middle of the centry.[186] The nobility and the higher officialdom was no longer the only body of persons banded together in the exercise of collective power. As the unprivileged too became capable of concerted action, the aristocracy renounced their traditional role as advocates of the free constitution, and supported the monarchical *coup d'état* of 1772.[187] But in 1789 the precarious balance was broken by the introduction of monarchic absolutism, after a revival of constitutional opposition from the aristocracy. This was accomplished through a coalition between the Crown and the non-noble interests in the lower Estates, whereby some essential noble privileges were destroyed to the benefit of the commercial bourgeoisie: social exclusiveness decreased along with the declining political influence of the first Estate; non-noble professionals flooded into the civil service as appointments were to be made upon merit rather than birth; the aristocracy lost parts of their land and some of their landed privi-

[183] See for instance Nils Andrén, *Svensk statskunskap,* Sth. 1961.

[184] Cf. Michael Roberts, 'Sweden', in A. Goodwin (ed.), *The European Nobility in the Eighteenth Century,* Black, London 1953, p. 145.

[185] Ibid., p. 146.

[186] The peasant rising of the Dalesmen in 1743 implied clear signs of a preference for popular absolutism on the Danish model rather than for a formal liberty which advantaged only the upper classes. In the Assembly of Estates, the victory of the 'Caps' over the 'Hats' in 1765 was based on a petit-bourgeois coalition–including the peasantry–against the privileged nobility and bureaucracy. See also chapters 8 and 9 below.

[187] King Gustav III broke the constitutional dominance of the Assembly of the Estates, to restore the old balance between the Crown and the Estates.

leges.[188] Royal absolutism was again reversed by the constitution of 1809, which was drafted to re-establish the balance between the King and the representative assembly. It was initiated by the noble interests, and there were still noble privileges prevailing in Sweden's political and economic life. But they were increasingly challenged by the liberal bourgeois interests and–as we shall see in later chapters–by the peasantry.

The landless rural proletariat, crofters and labourers, formed only about 1/4 of the number of farmers by the middle of the 18th century, and even if they had doubled by 1815, they still constituted fewer persons than the landholding peasantry. Both the peasant farmers and the noble and bourgeois proprietors increasingly used crofters as a labour force, although it was the population growth from the 1820s onwards which decisively created a substantial rural proletariat.[189] (See chapter 6 below.)

If we look at the *relations of authority*, the Swedish peasantry were not personally dependent upon the lords of the land in the Danish manner. The relationship between the Swedish landowner and his tenant (the *landbo*)–whether the landowner was the crown, the church, a nobleman, or a bourgeois or peasant proprietor–was a free, contractual relationship without personal jurisdiction delegated to the landlord.[190] Compulsive residence systems of the Danish model never existed in Sweden, and the village regulations of mid-18th century were, as we have seen, initiated and controlled by the public authorities.[191] But the leaseholding tenants on noble land, the *frälsebönder*, had no guaranteed security of occupancy; they could be taxed at the landlord's discretion, and their labour duties were often unrestricted. The physiocratic theory of the late 18th century, with its ideal of complete farmer responsibility and proprietorship, had few real consequences for this system, even if the labour duties prevailing in the fertile district of Scania were changed into an annual money rent.[192] Even the land of the freeholders, the *skattehämman*, was a kind of mortgage for the royal tax, since no peasant could sell his farm outside

[188] Cf. Roberts in Goodwin (ed.) 1953, op.cit., p. 151; see also Carlsson 1949, op.cit., pp. 261 ff. The commercial boom reinforced the trend, and strengthened the non-noble merchants, the smaller gentry, the industrialists, and the bourgeois civil servants.

[189] Cf. Hechscher 1957, op.cit., pp. 192–193. The results of very detailed investigations are presented in Gustav Utterström, *Jordbrukets arbetare*, vol. I–II, 1957.
See also Nils Wohlin, *Den jordbruksidkande befolkningen i Sverige 1751–1900, Emigrationsutredningen, bilag 9, Sth. 1909.*

[190] *Cf. Lars-Arne Norborg*, 'Landbo', in *Kulturhistorisk Leksikon for nordisk middelalder*, op.cit., vol. X, p. 202.

[191] Cf. Jutikkala 1963, op.cit., pp. 262 ff.

[192] Ibid., pp. 247.

his family without an optional priority being given to the Crown.[193] However, during the attacks upon the noble privileges towards the end of the 18th century, even peasants were given the right to acquire non-manorial noble lands, which could then be taxed by the state.

In spite of the generally high formal autonomy in land occupancy and management of plots, the economic obligations of the 18th-century Swedish peasant seem rather heavy.[194] Like his Norwegian contemporary he paid a basic land rent, the 'ordinary rent', to the crown. Then he paid special dues to sustain the public judicial system and the military defence. He also paid a church tax, usually a tenth of his produce; he was subject to charges for the right of home distillation; and he had to contribute to the conveyance of travellers. Thus the agricultural surplus obtained by the peasantry constituted the direct economic backbone of the nation-state and its administrative services.The tenants on a manorial estate rendered to their lord all those dues and services which they would have rendered to the crown if they had been crown tenants or freeholders, and on non-manorial noble lands they rendered half to the landlord and half to the crown.[195] But, as we have seen, the proportion of *frälsebönder*, whose relationships to the wider societal order were mediated by noble lords, was decreasing.

The inheritance system was quite complicated and varied according to the different type of *jordnatur* ('land type')–noble land, crown land, or taxable peasant land. On the crown land the division of the inheritance included only the personal chattels of the testator, while the farmer himself determined the future use of the farm, provided the district governor did not intervene.[196] The noble landlords were eager to prevent partition of their farmland, and controlled the order of succession. Even on the taxpaying peasant land the public authorities tried to limit the subdivision of farms in order to keep the tax subjects undivided. This coincided with the traditional inheritance system, according to which the eldest son usually took over the farm and bought out his co-inheritors at a reasonable price. But from the middle of the 18th century the public resistance to subdivision (*hemmansklyvning*) was revised, and the population growth also favoured the establishment

193 Ibid., p. 246. The feudal element was legally confirmed in 1734.

194 Cf. Ingers, vol. II, op.cit., 'Bondens bördor och besvär. 1700-talet', pp. 183 ff.

195 Cf. Roberts in Goodwin (ed.) 1953, op.cit., p. 145. Still in 1772 the *frälsehemman* (noble farms) outside the manors and the villages where the manors were situated made up more than 2/3 of noble lands, while the manors constituted hardly 1/5 of the total amount of *frälsehemman*–in spite of the tendency towards large-scale manorial farming and the sale of scattered farmland. The nobility was thus still to a high degree an agrarian *rentier* class, although the nobles had become agriculturalists to a greater extent than before. Cf. Heckscher 1957, op.cit., p. 186–after revised material from the land registers.

196 Cf. Jutikkala 1963, op.cit., p. 359.

of minor farms or crofter plots for younger brothers and sisters on the parental farm.[197] A strong protective institution like the Norwegian allodial privilege, the *odelsrett*, did not exist in Sweden, although some corresponding devices prevailed, as we shall see below.

To sum up, we have demonstrated the close interconnection between social hierarchy, authority relations, economic obligations, and institutional order.

In Norway the peasantry was not confronted with a landed nobility, with rural lords above them, but with an elite controlling the nationwide economic and political networks of society: the commercial bourgeoisie controlled market exchange, and was partly fused with the authoritative public bureaucracy. In the local setting then, the peasantry had a relatively high degree of land control and personal independence, while the major threat to their adaptation was the tax-collecting state. Their principal institutional defence mechanism, to resist the intrusion of market relations that might transform the possession of land, was the old allodial privilege.

The Danish peasantry lived within the same macro-political setting as the Norwegians–an absolute monarchy with no political organization into Estates–but the landowning upper class was the dominant mediation group between the peasant communities and the national order. Through patrimonial as well as mercantile domain they had a strategic broker function between officialdom and the peasantry, even after the weakening of the old families. Institutional defences like the Norwegian *odelsrett* were incredible, and would indeed have been totally meaningless, in this system.

Sweden was an estate society in a legal sense, with the privileged nobility as the top layer in society both administratively and in the countryside. But the nobility seems to have dominated the peasantry as a regulating and tax-collecting official agent rather than through partimonial domain in the local setting, and left the peasants free of personal dominance on the Danish model. The presence of rural lords as well as the remnants of feudal jurisdiction left, on the other hand, no room for allodial privileges of the strong Norwegian type.

'The old peasant society'

How do the aspects of hierarchical relationships discussed above relate to the land tenure systems in a more extensive conception of agrarian structure?

Integrated village settlement of the Danish type, controlled by a manorial estate, seems to be a precondition for duties to be rendered

[197] Cf. Heckscher 1957, op.cit., pp. 184 f.

in labour. The nucleated village is easily controlled through compulsive residence by powerful landlords, and results in an available labour force on a quite large and versatile scale. In Denmark the tendency to establish unified manorial estates producing for the market was also based on obligatory labour duties from the peasantry.

In many parts of Sweden the proportion of single farms outside the village system was greater, and the institutional integration of villages controlled by local lords generally weaker. Here the rent to the landowner was paid in kind or in money rather than in labour duties.

In Norway, where villages were altogether lacking, labour duties were extremely rare. They existed only in a few instances very close to manorial sites, but corresponding burdens were, on the other hand, laid upon the proletarian crofters living within the household domain of the rich peasant proprietors in the eastern and central districts. Transport duties were also imposed on the peasantry close to mining centres. Tax and rent were generally paid in produce, and the relative independence of the peasantry was reinforced by institutional defence –the *odelsrett* expressing kinship domain–and based upon the lack of landed lords. Norwegian agricultural lands were, as alway, meagre and relatively unattractive.

The Norwegian peasant society around 1800 was based upon the single or, more often, the multiple farm. The farm was an economic unit oriented towards subsistence production, and the peasant farmer was not a specialized professional, but a member of a versatile social group. Within the farm community there was a certain specialization in tasks and crafts, and most of the tools and articles for everyday use were produced and constructed on the farm or in the local community. The neighbourhood community consisted of a relatively isolated cluster of farms which were tied together through social intercourse and voluntary communal work, with a high degree of co-operation and fellowship.[198] The farms needed a certain product surplus for taxes and exchange, which was usually provided for through part time and secondary occupations like forestry, fishing, handicrafts, or labour services.[199]

In some respects Norway appeared as a dualistic society by the time we are considering.[200] The urban society, with a rich upper class of merchants and officials and a numerous lower class of servants, workers and beggars, was a capitalistic society, based on money economy, market exchange and profitable investment. The peasant society showed a markedly different structure of social organization. The rela-

198 Basic institutions in these closely knit communities were the *dugnad,* the exchange of mutual help when bigger jobs had to be done, and the ordinary *bytesarbeid,* the 'exchange work' in daily life–cf. Rigmor Frimanslund, *SEHR* 1956, op.cit., p. 72 ff.

199 Cf. Steen 'Bisysler . . .', 1933, op.cit., pp. 250 ff.

200 Cf. Steen, *NFLH,* vol. VII, op.cit., pp. 254 ff.

tively isolated peasant households and communes were patriarchal and partly self-sufficient, with an upper layer of peasant proprietors and prosperous tenants and a lower class of higly dependent servants and crofters. This society was characterized by a non-capitalistic peasant economy, a family household economy producing for subsistence, even if processes leading to the disintegration of this character were under way. The public demands of taxes and services, the increasing requirements of duties to be rendered in money, the military service and the wage-labour market advanced the integration of the peasant communities into the wider societal network. Thus the commercial penetration, the intrusion of a money and market economy, was a major transforming element introduced into the old social pattern. The dual society character does not then mean that the urban world and the peasant order were unrelated to each other. On the contrary: the peasant society was subordinated to the public authorithy and the privileged urban commercials in a power relationship implying a claim on the economic surplus. But the peasant society was characterized by a mode of production alien to the urban market economy, and by a social organization adapted to family domain and communal integration. The following chapters will be a specification of the transforming forces.

It follows from what has been said earlier that this picture of the Norwegian peasant society would be somewhat misleading when applied to Denmark and Sweden. The patriarchal figure in the Danish village was not the head of the extended family group, but the manorial proprietor, or his deputy in the case of absentee landlordism. The 'dualistic' element here was confined to the socio-cultural peculiarity of village life.[201] The landlord did not incarnate a social and economic organization alien to the authority and principles of the national order, but he was the mediating expression of this order at the local level. The transformation of the Danish peasantry was thus no intrusion of forces from outside, but rather internal to the structure of the agrarian society as such, where production for the market was taken up by the dominant class of landowners. The situation of the Swedish peasantry varied widely between geographical areas and 'types of land', but generally the peasants were faced with a privileged nobility within a semi-feudal order. The local lords were also an official elite in political and administrative terms, but economically contradictory to the new commercial interests which questioned their formal prerogatives. Broadly speaking, Sweden was thus not only split between a peasant sector and a wider commercialized society, but between noble privileges and rising bourgeois demands within the national order itself.

[201] Cf. Joan Rockwell, 'The Danish Peasant Village', *The Journal of Peasant Studies*, Vol. 1, No. 4 (July 1974), pp. 409–461.

CHAPTER 6

The transformation of landholding and tenure systems

The basic changes in landholding and social order which occurred during–roughly–the late 18th and early 19th centuries–with important regional variations beyond this period–might be summarized as an *individualization of agriculture*. In relation to rural power groups outside the peasantry, their subordinate land disposal was gradually replaced by widespread peasant proprietorship even in former landlord areas, and the close peasant communities were broken up by enclosures and individual landholding.

Both these processes transformed the relation to land as a factor of production, as they led to greater possessiveness and scope for manoeuvre. Contractual relations of society gradually permeated rural life, and made market integration structurally feasible, although the process was far from accomplished all over the Scandinavian area during the period in question.

Separate ownership was necessary for the economic control of one's own estate, and so were enclosures of communal land and reallotment of intermixed strip holding. But agricultural individualization also implied greater social differentiation, since the dissolution of closely knit communities introduced clear-cut hierarchies within peasant society itself, based on the formal possession, or lack of possession, of consolidated lands.

Even so, the rather similar processes were not identical. They occurred in highly different ranges and sequences in each of the Scandinavian countries as well as regionally within them, and their socioeconomic meaning varied according to the contextual pattern.

We will first trace the development towards peasant freehold in each of the three countries. The comparative pattern will be hinged on the questions *when* the transition occurred, *how* it happened initially, *what* the process implied in terms of landholding quality and social differentiation, and *why* the transformation basically took place: which were the main causal factors as they might tentatively be isolated in the analysis of separate sequences and similar or dissimilar features.

Then the enclosure movement and reallotment of communal lands will be compared, and, finally, discussed in relation to the move towards freehold.

The transition towards peasant proprietorship

The Norwegian case

The decisive move from tenancy towards freehold peasant ownership in Norway had already started in the last decades of the 17th century. Earlier, areas of predominant peasant proprietorship had been restricted to the mountainous districts in the south, from Agder to the interior of Telemark, and to some remote parts of the eastern and western valleys. But by the middle of the 18th century, it is generally held that the freeholders were in a majority in the country as a whole.[1]

According to *Sars* and his successors, the tenants outnumbered the freeholders by 3 to 1 just before 1660, while in 1750, on the contrary, the freeholders were twice as numerous as the tenants.[2] But the proprietary rights of the freeholders were not so unrestricted as held by Sars, since some of them possessed only a part of the rent value of the farm, and the distinction between tenants and freeholders was, on the whole, less easily made. The system of land ownership was, as we saw in the preceding chapter, based on rent-property, and it has been documented that the peasantry possessed scarcely 1/3 of the total rent about 1660.[3]

It is estimated that the share of the freeholders was nearly 2/3 of the land in 1814, and this proportion had already been reached by the middle of the 18th century.[4] Thus the first long wave in the transition towards peasant proprietorship flowed from 1660 to 1750. This covered mainly the broad wooded valleys and plain areas in the eastern and south-eastern parts of the country.[5] Still about 1820 1/6 of the registered rent was public property, mainly ecclesiastical estates connected to clerical offices, cathedral schools or semi-public foundations, while about 1/5 belonged to bourgeois proprietors.[6] In the south and in the eastern mountain regions more than 90 % was peasant property, compared to about 70 % on the eastern plain and valley area,

1 See, inter alia, Sars, *Samlede Værker,* vol. 3, Kristiania and Copenhagen, 1912, pp. 148–153, and Skappel, 'Det norske jordleievesens geografiske utbredelse i eldre tid og overgangen fra leie til selveie', *Historisk Tidsskrift,* vol. 30, Oslo 1934–36, p. 240 (cf. also note 46 to chapter 5.).

2 Ibid.

3 See Holmsen, *SEHR,* IX, 1961, table on p. 162; cf. chapter 5, note 50.

4 Ivar Mortensson, *Bondeskipnad i Norig i eldre tid,* Kristiania 1904, p. 9.

5 Cf. Sars, *Samlede Værker,* vol. 3, op.cit., pp. 152–153; Holmsen, *SEHR,* 1961, op. cit.; Gudmund Sandvik, 'Jordleigar og jordeigar i norsk historie', stencilled, 1969.

6 Bjørkvik & Holmsen, in *Heimen,* vol. IX (1952–54), op.cit., p. 83.

about 65 % in Trøndelag, a good 60 % in the western districts, and only about 10 % in northern Norway.[7] But the move towards peasant proprietorship started anew in the 19th century. While about 1/3 of the total registered rent comprised estates let on lease in 1820, this was reduced to a good fourth in 1840, scarcely 1/7 in 1860, and scarcely 1/10 in 1890.[8] This move affected nearly all the agricultural lands of central and southern Norway, where the freeholders in 1835 constituted more than 3/4 of the peasants in all districts, except in the outer coastal area in the west, where they still comprised only 55 %.[9] But in northern Norway the percentage of freeholders was only 20 % in 1835, and here the transition towards peasant proprietorship did not occur until the last decades of the 19th century.[10]

The distribution between freeholders, tenants and lot-holding crofters in the different counties in 1835–based on the census data–has been estimated as follows:

[7] Ibid.

[8] Simen Skappel, *Eiendomsbesiddere i Norges Landdistrikter*, Kristiania 1918, p. 15.

[9] Cf. Holmsen, *SEHR*, 1961, op.cit., p. 162. His table here compares the amount of peasant property in 1661 with the percentage of freeholders in 1835 for the following regions:

Peasant property in Norway in 1661 and freeholders in 1835

	Peasant property 1661 as % of total rent*	Peasants (farmers) 1665	Peasants (farmers) 1835	% age of free-holders
1. Northern Norway	8	c. 5,000	8,905	20
2. Trøndelag	10	5,000	7,368	77
3. Outer Western Norway ..	15	11,000	19,845	55
4. Inner Western Norway ..	31	7,500	13,134	72
5. Agder (Southern Norway)	76	5,000	10,314	89
6. Eastern mountain communities	83	2,000	6,245	88
7. Eastern valley communities	49	3,300	8,644	89
8. Opplandene	33	4,500	9,557	85
9. Oslo and Skiensfjord region (excl. Larvik)	30	7,000	16,788	80
10. Earldom of Larvik	24	700	1,607	40
Total	31	c. 51,000	102,416	70

* Based on tables compiled by H. Bjørkvik and A. Holmsen in Heimen, Vol. IX, (1952–54).

The percentages in columns 1 and 4 are obviously not directly comparable, as the former relates to land property (assessed by *landskyld*) and the latter to landowners. The difference between them is nevertheless a relevant index of the progress of peasant freehold ownership from 1661 to 1835.

[10] Ibid.

	County	Free-holders	Tenant farmers*	Crofters with plots of land
	Smålene	4423	1209	3277
	Akershus	4152	1523	4435
East	Hedemarken	5962	941	4990
	Christian	6128	625	7788
	Buskerud	5982	856	3732
	Jarlsberg & Laurvig	3950	1489	1007
	Bratsberg	4245	1365	3891
South	Nedenes og Råbygdelag	3829	521	1617
and	Lister & Mandal	6131	819	695
south-	Stavanger	4311	1659	3088
west	Søndre Bergenhus	5856	3830	3354
	Nordre Bergenhus	3824	2431	3512
West	Romsdal	3860	3255	3154
Centre	Søndre Trondhjem	4749	1458	4292
	Nordre Trondhjem	2697	1431	4640
North	Nordland	1096	5029	1113
	Finnmarken	1429	2127	628
Total		72624	30568	55213

Source: A. M. Schweigaard, *Norges Statistik*, Christiania 1840, p. 35.

* The tenant farmers include everybody using registered land that they don't own themselves–like clergymen, civilian and military officials, and church attendants, whose offices entailed the possession of land. But these comprise hardly 4 % of the stated number of tenants, according to Schweigaard 1840, p. 54.

How did this transformation of land disposal occur? The twin kingdom had an enormous national debt at the introduction of absolute monarchy in 1660, mainly due to the war of 1657–60.[11] The general expansion of state activity during the 17th century also meant a growing need of financial means. In the years from 1660 to 1675 there was a substantial sale of crown lands, especially big estates, to supply the public purse. The estates were mainly sold in bulk to rich creditors–partly to foreign capitalists who had supplied the state with commodities and loans, and partly to civil servants, officers, and noble bourgeois proprietors who had some pay due.[12]

[11] The debt amounted to about 5 mill. *riksdaler*–cf. O. A. Johnsen. 'Fra leilænding til selveier', *Nordisk Tidskrift*, Stockholm 1910, p. 278.

[12] Cf. Johnsen 1910, op.cit., p. 279. From 1660 to 1675 Crown property was sold for nearly 1.3 mill. *riksdaler*; cf. also Skappel in *Historisk Tidsskrift* 1934–36, op.cit., p. 239.

The first transference of crown lands came to a standstill during, and immediately after, the new war against Sweden in 1675–79. These original sales did not advance peasant proprietorship, and were rather a drawback to the leaseholding peasantry. The new absentee landlords often exacted illegal charges, and increased the land rent and the leasehold duties, to get bigger returns from their invested capital. The peasants saw it as a change for the worse to become dependents of private proprietors instead of tenant farmers on crown land, and there was a quite widespread demand, expressed in a lot of written complaints referring to the exorbitant profits of the proprietors, that the king should take back the lands which he had disposed of.[13] But from the end of the 1680s the system of land-rent openly decayed, and the exceedingly important move towards peasant proprietorship took place. The sale of crown land started anew, and now a substantial number of farms were sold directly to the peasants. A substantial part of the land belonging to public officials and noble and bourgeois proprietors was also passed on, by sale to the peasantry, particularly in the eastern plains and forest area.[14] The sale of crown land then stopped in the 1690s, but started once more around 1720. This time only scattered crown property, *strøgods,* and church estates were sold, since the monarchic ruler claimed to be the lord of the church. In a majority of cases the best farms were bought by burghers and civil servants, but a substantial number of smaller farms were also sold direct to peasant farmers, not only in the east, but also in southern and western districts.[15] The lands were sold with a right of redemption reserved to the crown, being in fact something between a sale and a loan, and the rise in prices towards the 1740s made the sale of this right a new public business during that decade.[16] Still a lot of peasants in the eastern districts were able to buy their single farms from the bigger proprietors.

The second great period in the transition towards freehold, in the 19th century, seems concomitant with the sale of ecclesiastical property, *beneficert gods,* with which the parish priests and some clerical foundations were endowed. A statute of 20 August 1821 provided that sale of this remaining public property was to be preferred to leasing or renting, and the government was instructed to sell crown lands to

[13] A résumé of letters of complaint, which were summarized by the Norwegian governor's office, the *stattholderkollegium*, and sent to the King in January 1672, is referred to by Johnsen 1910, op.cit., pp. 282–283.

[14] Mainly in the counties of Hedmark, Opplandene, and Akershus–cf. Skappel 1934–36, op.cit., p. 240. See also Johnsen 1910, op.cit., p. 285.

[15] Cf. Steen, *NFLH*, vol. VI, p. 88.

[16] The reservation of the right to buy the Crown estates back was sold for about 1/3 of the original selling price–cf. Coldevin, 'Enevoldstiden', *Vårt folks historie*, vol. V, Aschehoug, Oslo 1963, pp. 199 ff.

private individuals when leases fell vacant.[17] This opportunity for the leaseholders to acquire the status of a freeholding peasantry nearly completed the transition towards peasant proprietorship in the agricultural districts of Norway. The only districts that were, on the whole, unaffected by the sales of crown land and private estates were the fishing communes of northern Norway. Here the part-time peasant fishermen did not acquire freehold to their land until the last decades of the 19th century, when commercial privileges were abolished and new credit systems introduced as we shall see below.

Let us see what this transformation of ownership relations implied for the character of land ownership as such, and for the social structure of the peasantry.

The transition towards freehold was stimulated by changing economic conditions, and the new peasant proprietor was not identical to the traditional *odelsbonde.* The tenants often bought their farms on a mortgage, and the mortgage bond, *pantobligasjon,* originated as a financial institution to replace the old mortgage deed, the *pantebrev,* in the beginning of the 18th century.[18] This meant that the capital was detached from farm produce and land prices, and attached to monetary value, since the peasant farmers had to pay interest instead of land rent. Since commercial goods increasingly permeated rural life, they also became indebted to the privileged marchants, especially after the bad crop failures of 1740–42, and thus gradually shifted their dependence from the landed proprietors to the urban bourgeoisie.[19] The new financial conditions meant faster trade and business in land, and in certain respects the new peasant proprietors, who bought their farm on a mortgage, were more loosely attached to the soil than the old allodial peasantry had been.[20] Partially, however, the relations between man and land–or better, between family group and land–were cemented by institutional integration through acquisition of the allodial privilege.

The freeholder with a mortgage debt often paid an annual interest that exceeded the old land rent and the leasehold duties.[21] But the

17 The Crown forests, including settlements, meadows, outfarm pastures, and allotments, were not to be sold, but the exception as to settlements and forest allotments was annulled in a statute of 1827, which also settled rules relating to surreptitious sale to leaseholders and their descendants; and from 1848 even the Crown forests themselves could be sold–cf. G. Tank, 'Salg av statsgods efter 1814', *Historisk Tidsskrift,* 5. R. 6.B., Oslo 1927, pp. 46–87.

18 Cf. Holmsen, 'Problemer i norsk jordeiendomshistorie', *Gard, Bygd, Rike,* 1966, p. 96.

19 Cf. Koht, 'Bonde mot borgar i nynorsk historie', *Historisk Tidsskrift,* V.R.1, Oslo 1910, pp. 29–85; see also Steen, *NFLH,* vol. VI, pp. 207–208.

20 Cf. Holmsen,''Gårdene' i den gamle matrikkel', *Gard, Bygd, Rike,* 1966, pp. 135–160.

21 Holmsen gives a concrete example from the district of Romerike, in *Gard, Bygd, Rike,* 1966, p. 97, note 14.

growing demands for agricultural products through the expansion of mining and forestry, implied potential profits through the expansion of commercial markets, and made long-sighted improvements of the farm, for the benefit of the heirs, economically meaningful. This was the general *rationale* of freehold. But still the new peasant proprietor in the eastern plains and broad valley area–often quite well-to-do with a lot of crofter plots on his land–seems closer to his predecessor, the rich tenant with partial rent-property in many farms, than he is to the traditional freeholding *odelsbonde* in the meagre and isolated mountain districts.[22]

We should then take a closer look at the crofter system, which seems rather intrinsic to the great transformation itself (cf. the institutional description in chapter 5). The growth of crofter plots and cottagers with labour duties on the peasant farm closely follows the transition towards freehold. The crofters are more numerous in those districts where the new peasant proprietorship is more original and widespread. If we group the counties of the eastern plain and wide valley area together, the development of crofters with plots of land runs as follows:

Landholding crofters[23]

	Absolute figures			Relative figures		
	Eastern plain & valley area*	Rest of country	Total	Eastern plain & valley area	Rest of county	Total
1723	8475	3339	11814	72	28	100
1801	24786	14625	39411	63	37	100
1825	27883	20688	48571	57	43	100
1855	32766	32294	65060	50	50	100

Source: Simen Skappel, *Om husmandsvaesenet i Norge. Dets oprindelse og utvikling,* Kristiania 1922, p. 168.

* Includes the counties of Østfold, Akershus, Hedmark, Oppland, Buskerud, Vestfold and Telemark.

These figures indicate clearly that the crofters were far more numerous in the eastern districts during the 18th century. This area showed, as we have seen, a marked move towards peasant freehold as early as the end of the 17th century. But the relative as well as the absolute increase in

[22] Cf. Holmsen, *SEHR*, 1961, op.cit., p. 164.

[23] Comprise married men over 20 years of age, and unmarried men over 30 years– Skappel, *Om husmannsvæsenet i Norge,* Kristiania 1922, p. 170.

the number of crofter plots was greater in the rest of the country during the first half of the 19th century, parallel to the later development of peasant proprietorship in these districts. While the number of land-holding crofters increased by 32.2 % in the seven eastern counties from 1801 to 1855, the corresponding increase was 120.8 % in the remainder area. The growth of crofter plots in the east probably slowed down by the middle of the 18th century, and the low increase during the first half of the 19th century meant that the growing demand for agricultural labour was satisfied by more labour duties from the older crofter plots. The labourer-crofters often got labour duties throughout the year on farms where they had been obliged to do only seasonal work before.[24]

In 1801 there was a total number of about 48 000 crofters–40 000 with plots of land and 8 000 without land–compared to about 90 000 peasant farmers (both tenants and freeholders), 8 900 day labourers, 40 000 servants, and 13 000 rural lodgers (mostly poor and lonely old people.)[25] From 1801 to 1855 the number of crofters increased by 81.7 %, the number of peasants by 25.9 % (mainly due to farm partition), day labourers by 225.3 %, and rural servants by 34 %.[26] After 1855, however, the number of crofters began to decline, both relatively and in absolute terms. The crofter class comprised about 26 % of the rural population in 1845, about 15 % in 1875, a good 4 % in 1910, and scarcely 3 % in 1920.[27] The absolute number of lot-holding male crofters during the 19th century was as follows:[28]

1801	1855	1875	1890	1910	1920
39411	65060	52769	29623	19763	14666

The crofter system culminated by the middle of the century and was, on the whole, under abolition from about 1870, when labour duties were

24 According to Skappel 1922, op.cit., p. 169.

25 'Crofters with plots of land' include married males over 20 years and unmarried males over 30 years; 'peasant farmers' include men over 20 years; 'servants' comprise all males minus 1½ times the number of all between 1 and 10 years of age; 'day labourers', count all men over 20 plus 1/2 of all in the category between 16 and 26 years; while the 'lodgers' are all males over 20 years. Cf. Skappel, 1922, op.cit., p. 170.

26 Skappel, op.cit., table on p. 170, compiled according to the official census statistics. The vast relative growth of day labourers should be seen against the background of a very low figure in 1801.

27 Skappel, op.cit., p. 177.

28 It should be noted that the classification was not identical from one census to another. Gradually more importance was attached to occupation, and some lotholders who were earlier called *husmenn* (crofters) were later entered into other categories. But this error cannot easily be corrected, since it also reflects a societal evolution.

gradually reduced and crofter plots shut down. The only exception was the three northern counties, where the number of crofters *increased* with the population growth after 1855, parallel to the transition towards peasant freehold in this area.[29]

The crofter system expanded with the population surplus of the 18th and early 19th century, when the desire of the rural proletariate for a house and a family met with the labour requirements of the economically expansive peasant proprietorship.[30] The system dissolved together with the closed peasant communities, when a majority of the landless perceived better alternatives–either as independent urban workers, or as immigrants to northern America, or as small-holding proprietors of their former crofter plots. The last option opened up from the beginning of the 20th century, and marks the final stage in the transition towards peasant freehold in Norway, when the crofters were stimulated to buy their own small lots and make a living as part-time worker-peasants. But this process is outside the schope of the present study.

How should the transition towards peasant freehold be explained? Why did the sequence of the process occur the way it did? Some suggestions can be made.

The transition towards peasant proprietorship in Norway was initiated by the extensive sale of crown lands after 1660, caused by the new financial needs of monarchic absolutism. This is generally held to be the starting point, but it can never have been decisive, since the state property was not originally sold to peasants, but to officials and bourgeois creditors. Why did the rich proprietors afterwards sell their estates in part to the peasantry? How could the leaseholders, in a subsistence agricultural economy, be able and willing to buy? The answer to these questions has been a controversial point in Norwegian historiography. *Sars,* in his original exposition, stressed the legal restrictions on the opportunities of the landlord to profit from estates let on lease.[31] By a decree of 2 February 1685, a double tax burden was imposed on leasehold property which was attached to the home farm. Thus it became unprofitable to possess landed estates outside the farm that was used by the owner himself. Sars also stressed that increasing prices for land made it profitable to sell, and that buyers could be found if the estates let on lease were parcelled out to several owners who farmed their land themselves.[32] The effect of legislation by the King, in favour of a freeholding peasantry, has particularly been

[29] See Skappel, 1922, op.cit., table on p. 170.

[30] Cf. Steen, *NFLH,* vol. VI, pp. 197 ff.; and Johnsen *Norges Bønder,* Kristiania 1919, pp. 272 ff.

[31] Sars, *Samlede Værker,* op.cit., vol. II, p. 389.

[32] Ibid.

stressed by O. A. *Johnsen*.[33] Several statutes of 1684–65 declared not only that a double tax should be imposed on all landed property additional to the home farm, but also that land rent and leasehold duties should be frozen; that leasehold should be guaranteed for lifetime; that civil servants were forbidden to buy land in their own district; and that the tenant should not be deprived of the main accompanying amenities. This legislation was obviously not effectively carried out in practice–probably it was, for instance, unrealistic to freeze land rent and leasehold duties in a period of inflation–but Johnsen argues that landed estates were made substantially less profitable in Norway, and thus that the social and economic independence of the Norwegian peasantry was intentionally increased. The motives of the King, and of the Danish vice-regent in Norway, were supposed to have been stimulated by the effectiveness of the army of free peasants in the war against Sweden in 1675–79, compared to the inefficient Danish mercenaries and conscript serfs.'[34] The other side of the coin–that the legislation, if effective, would check the development of a strong and independent upper class in Norway–was probably just as important.

The opposite school of thought took the ineffectiveness of the legislation as its point of departure, and stressed the general trend in prices for land and agricultural produce as a causal factor. *Koht* argued that while prices steadily increased throughout the 16th and 17th centuries, with a subsequent demand for land as a profitable investment of capital, this increase came to a standstill–or prices even began to fall–about 1670.[35] Then the capital withdrew from agriculture, and found more profitable investment in mining and other kinds of trade and industry, while landed property did not give a sufficient rent. When land thus became cheaper, the peasants got an opportunity to buy, and through the substantial sale of Crown land, the general price of land was pressed down even further.[36]

[33] Johnsen, 'Fra leilænding til selveier', *Nordisk Tidskrift,* Stockholm 1910, pp. 283 ff., and Johnsen, *Norges Bønder,* Kristiania 1919, pp. 271 f.

[34] Ibid.

[35] Koht, *Norsk Bondereising*, Oslo 1926, pp. 198 ff., and Koht, 'Prisar og politikk i norsk historie', *Samtiden* 1913, p. 249.

[36] *Steen* also stresses the economic conditions in Europe in the 17th century, and the price depression on agricultural goods which made landed estates relatively less profitable–cf. *NFLH*, vol. V, pp. 262 ff. *Koht* also supplied his thesis with a political argument, stressing the lack of opportunity to meet the depression by squeezing the peasantry further (which was done in Denmark and continental Europe), owing to the free position and traditional resistance of the Norwegian peasants. His stubborn insistence on the price cycle as a general formula for political explanation seems rather idiosyncratic–'the high prices create a spirit of rebellion, and the low prices give the strength of victory', *Samtiden* 1913, op.cit., p. 252. His price material seems, moreover, scanty, and gives no convincing explanation of the complete economic conditions of a subsistence-oriented peasantry.
See also Stein Tveite, *Bønder og prisskifte*, University of Oslo, 1953.

These explanatory factors seem too general and unqualified to account for the specific sequence and pattern of the transition towards freehold in Norway. They leave the uneven development and regional variations unrelated to the structural characteristics of agriculture and land tenure. Firstly, it is obviously right that many of the bourgeois proprietors found alternative capital investment, in mining and lumber export, more profitable than the possession of landed estates let on lease. This indicates the fundamental weakness of the system of tenancy in Norway. The peasantry were mainly producing for subsistence, with most of their monetary income from secondary occupations outside agriculture, and they had no stimulus to improvement or even proper maintenance of the farm. The agricultural conditions were generally bad, and did not stimulate the new capitalism to sink its investments in land.[37] Secondly, the proprietors got much higher prices when they sold to peasants than they had paid for the crown land in the first place. In addition, they got a mortgage in the farm, and a stable interest.

But why were the peasants motivated to buy, and how could they afford to do so? As we have seen, the Norwegian peasantry, also as leaseholders, were personally free, without any pressure from adscription or villeinage. They mainly worked a subsistence agriculture, where tenancy in itself was no hindrance; and the land rent and leasehold duties were on no account heavier than the purchase price and the interest on a mortgage bond. Still a lot of peasants were able and willing to buy their own farms. The proprietors had bought their estates from the crown lands in the first decades after 1660 at prices based on capitalization of the rents. But poor agricultural maintenance of the leasehold property implied that the farm value based on the rent income decreased relative to registered tax assessment. The peasants, however, were able to pay many times the price based on capitalized leasehold rents, provided they used the farms themselves. The peasant family farm is not, as we saw in chapter 2, a capitalist enterprise with farmland as one among alternative investment options, even if this was the case for the bourgeois proprietor, who had no immediate subsistence motivation for his investment in land. But this difference is only a basic prerequisite, which is unable to explain the preference of freehold to tenancy. The choice of freehold looked beyond the mere subsistence orientation, and meant a prospect of future profits which compensated the high price. Proprietorship was a prerequisite for the prize of farm improvements to be won by the peasant farmer himself, his family and his heirs. Farm improvement required a certain reinvestment of surplus, which implied some market production in order to obtain more advanced equipment. Alternatively, ownership of farm

[37] See Skappel 1934–36, op.cit., pp. 244 ff.

land could be the means to exploit subsidiary riches. Thus the internal contradictions in the system of leasehold and rent-property gradually made it an unprofitable business for the owner as well as for the leaseholder, so that the peasantry could now buy their own land against a mortgage bond.

What about the regional variations? Norwegian agriculture hardly developed beyond the non-market subsistence level during the 18th century, and the interest in freehold seems limited to those districts where forests were part of the farm land. There was a vast increase in profitable timber commerce from the 17th century, and the original transition towards freehold is strikingly predominant in the wooded eastern regions of the country.[38] This is also the area where the crofter system first developed to supply the freeholders with a flexible labour force in land clearing and forestry. The urban commercial class, who had a monopoly in timber export and sawmill operations, kept vast forest areas for themselves, but they also made profitable contracts with the freeholding peasantry, to get timber effectively from peasant ground. The first move towards freehold seems stimulated by mercantilization of goods outside agriculture itself, while the non-commercial subsistence character of farming was preserved. When peasant proprietorship spread to the remainder area in the 19th century, this could be interpreted as a response to increasing mercantilization of agriculture as such.[39] This hypothesis will be further illuminated by the next chapter, but it is closely in accordance with our model of the agrarian structure in Norway, with its lack of dynamic landlordism. The process of commercialization seems to be a fruitful explanation of the pattern and sequence of increasing freehold, although the causal relations are complicated and obscure–the interconnection belongs to a specific structural context.

The fishing districts of northern Norway already spoil this neat picture, since commercial exports were basic–centuries before freehold was introduced. The sale of Crown lands was no germ to peasant freehold along the outer coastal area and northwards, but led to permanent bourgeois estates based on the commercial profits from fishing, whaling and sealing, especially in northern Norway. The modest land rent meant little compared to the nesting places, fishing grounds and stations.[40] To the peasant fishermen, fishing was a hazardous task, and so long as they were deeply in debt to the fish export monopolists in Bergen, they saw no advantage in freehold. If they remained tenants (and they usually stayed safe on their farms for generations), the cre-

[38] This connection is especially emphasized by Holmsen, *SEHR*, 1961, op.cit., and by Sandvik 1969, op.cit.

[39] Cf. Sandvik, op.cit.

[40] According to Skappel 1934–36, op.cit., pp. 251 ff.

ditor was motivated to renew the working capital for continued fishing, whereas he could take the farm if the debtor was a freeholder. Agriculture was secondary in the trilateral system between fisherman, merchant and landowner. The labour surplus was extracted by the merchant as profitable goods for sale, and by the landlord as land rent and leasehold duties, while the peasant fisherman, in return, was supplied with working equipment and credit from the one, and a small farm with contractual use, independent of the commercial debt, from the other.[41] This was a strongly patriarchal system, with very little strategical leeway for the peasantry. The transition towards peasant proprietorship did not start until the second half of the 19th century, when liberal legislation broke down the tight commercial monopoly in Bergen, when steam ship routes made transport easier, and when the new banking system replaced the credit function and financial dominance of the commercial class.[42] This institutional change deprived the system of tenancy of its foundations, and some proprietors also sold their land when the sea was declared free ground in a statute of 1857, to the disadvantage of the *vœreiere,* the lords of the fishing stations.[43]

Towards Danish freeholds

The transition towards peasant freehold in Denmark started carefully in western Jutland with sporadic purchases in the 1760s. Some manorial estates were split up around 1760, and a few enterprising leaseholders bought their farms at the auction sales. The main manorial reform, however, was the tendency to replace ordinary leasehold by hereditary tenancy, or copyhold. This occurred outside the old area of villeinage– in Jutland and to some extent in Funen. In Zealand, on the contrary, son succeeded father on the farm not quite half so often.[44] This original move stagnated during the 1770s and 1780s, but there was a rapid development of peasant proprietorship during the decades around 1800 – declining towards the state bankruptcy in 1813.[45] Freehold and copyhold now became dominant in Jutland, and covered a substantial proportion of the land in Funen, but were still insignificant in Zealand.[46] The development stopped as a result of the credit squeeze brought

[41] Cf. Axel Coldevin, *Jordegods og storgårder i Nord-Norge,* Trondheim 1943, pp. 18–32.

[42] Ibid., pp. 68–76.

[43] Cf. Skappel 1934–36, p. 256. I have not been able to control the practical implications of this legal measure against empirical investigations, so the range of the argument should perhaps not be stretched too far.

[44] Sigurd Jensen, *Fra patriarkalisme til pengeøkonomi,* Copenhagen 1950, pp. 16 ff.; F. Skrubbeltrang, *SEHR* 1961, op.cit. (note 42 to ch. 5), pp. 167–68.

[45] Cf. Jensen 1950, op.cit., pp. 34 ff.

[46] Cf. Skrubbeltrang, *SEHR* 1961, op.cit., p. 165.

about by the public bankruptcy of 1813, and there was a further worsening of the condition of freeholders in the agrarian depression from the beginning of the 1820s. Several freeholders now had to abandon their farms, and some even returned to leasehold contracts.[47]

In 1835, the first year for which accurate figures are available, there were nearly 42 000 freehold farms–including 6000 with copyhold tenure–against nearly 25 000 ordinary leasehold farms.[48] The freeholders possessed 56 % of the *hartkorn* (the land rent value) against 44 % held on lease. Thus the freehold farms were, on the average, smaller than those let on lease–partly owing to greater partition of freehold property, and partly owing to concentration of freeholders in the *hartkorn*-deficient Jutland.[49] Freehold proper now included about 15 % of the farms in Zealand (where copyhold tenure prevailed in the northern Crown land area), about 40 % in Funen, somewhat less in Lolland-Falster, and more than 70 % in Jutland.[50]

The third period in the transition towards peasant freehold in Denmark started in the 1830s. At first purchases predominated in Jutland, as ever, but from the end of the 1840s, the process became particularly strong in Zealand and the other islands. This is reflected in the following figures:

Freehold in percentage of the rent value (hartkorn) *of Danish peasant farms*

Year	The islands	Jutland
1850	44 %	80 %
1860	59 %	89 %
1873	76 %	94 %
1905	89 %	99 %

Source: Skrubbeltrang 1938, p. 79 and p. 85.

About 1850 the total number of freehold farms (including the hereditary tenancy estates) was nearly double the number of leasehold farmsteads, but still they amounted to only a good half of the total *hartkorn* value, because farm partition and sale of parcels was the privilege of freeholders.[51] The decisive transition from tenancy to proprietorship

[47] Jensen 1950, op.cit., pp. 81–82.

[48] From the *Rigsdagstidende* 1871–72, Tillæg A. Sp. 1929–ref. by Jensen 1950, op.cit., p. 89.

[49] Cf. Skrubbeltrang, *Den Danske bonde 1788–1938,* Copenhagen 1938, p. 79.

[50] Skrubbeltrang, *SEHR* 1961, op.cit., pp. 173 ff.

[51] Jensen 1950, op.cit., p. 89.

in the old area of villeinage took place during the second half of the 19th century, although the old leasehold system was not formally abolished until 1919.[52]

Unlike the processes in Norway, there was no transition towards peasant proprietorship by the sale of Crown lands after 1660. The Crown estates were mainly bought in bulk by the new service nobility and the prosperous landlords, and virtually nothing was forwarded to freehold. There occurred some sales of Crown lands again in 1765–67, and now there were certain instances of peasant purchases at auctions, although most of the land was sold as unified estates, since financially strong buyers supplied more to the public purse. Some leasehold property was also transmitted to the tenants, and there was some peasant purchase of the rights to the accompanying amenities.[53] The government was opposed to estate partition and sale of leasehold property in the 1760s, but it stimulated the transition towards hereditary tenancy from the end of the decade.[54]

The decisive move towards peasant freehold–in the second period from the 1790s to about 1810–was directly stimulated by the state. Some of the purchases were supported by capital on loan from the new royal credit bank, and the great reform legislation, especially the restriction of compulsory labour service (the *hoveri*), made leasehold estates less attractive to the landlords. Peasant proprietorship was also furthered through 'estate liquidation' by speculators in real property, who bought landed estates only to parcel out the main fields and sell the leasehold lands against mortgage.[55]

In the third period–towards the middle of the 19th century–most of the freehold purchase was financed by private capital, partly through credit from the seller, where the landlord got a preferential claim, partly through loans from friends and neighbours, and partly through the purchaser's own capital, which, however, was often negligible. Despite the general development in mortgage banking, the state was reluctant to finance the purchase of freeholds for fear of inflation.[56]

The move towards peasant proprietorship in Denmark generally implied a transition from a patriarchal landlord-leaseholder relationship to a capitalist creditor-debtor relation. The net capital among the

[52] An act of 30 June 1919 stated that leaseholding peasants and crofters could obtain proprietorship on certain conditions. The peasant leaseholders by then had about 2 % of the *hartkorn*.

Cf. Skrubbeltrang 1938, op.cit., p. 91.

[53] As we have seen, this took place on a rather small scale, and was mainly limited to Jutland–cf. Jensen 1950, op.cit., pp. 22–23.

[54] The government codified the first purchases in a statute of 13 May 1769–cf. Jensen, op.cit., p. 32.

[55] Ibid., pp. 49 ff.

[56] Ibid., pp. 89 ff.

peasantry was negligible, and credit was supplied against a mortgage on the farm. But the transition towards non-personal, contractual relationships was a gradual one. Particularly the first purchases of freeholds were mediated by rich benefactors, like the parish priests, who often gave security.[57] During the first period, the way for later peasant proprietorship was also prepared by the development of tenancy conditions as a temporary alternative, through hereditary leasehold within paternalistic landlord control.[58]

The character of peasant freehold is conditioned by the dependence upon creditors. In many cases of peasant purchase, the landlord became the main creditor, and retained a certain patriarchal position through the local dominance of manorial mortgage. But many landed debenture holders also sold their bonds to others, and towards the end of the second period, about 1810, it is estimated that each single estate owner disposed of only a limited amount of preferential claims.[59] There were also a great number of lenders within each part of the country, and thus no real monopolies in the giving of credit to the peasantry. The peasant dependence upon creditors seems to have been increasingly fragmented by claims from landlords, officials, the commercial bourgeoisie, and the state–which probably afforded the freeholders some tactical mobility.

Later, in the 1880s, the new agrarian depression (from 1877) became particularly hard for the heavily indebted freeholders, since the interest was often worse than the old leasehold duties. While the deed of conveyance spoke of a free man on his own land, the mortgage bonds told another story.[60] The gradual transition from patronage and economic dependence to autonomous entrepreneurship meant new impersonal constraints.

The processes leading to 'peasant emancipation' and freehold also had important consequences for the crofter system.[61] About the year 1700 there had been more peasant farmers than crofters all over Denmark, ranging from 4 against 3 in Zealand, about 3 against 2 in Funen and the southern islands, and about 2 to 1 in Jutland.[62] But the number of crofter plots tripled throughout the 18th century, while the number of peasant farms stayed more or less the same. The most remarkable change occurred in the last decades of the 18th century. There was a rural population growth of about 60000 from 1787 to 1801, and the

57 Ibid., p. 30.

58 Ibid., pp. 28 f., cf. Skrubbeltrang, *SEHR* 1961, op.cit., p. 175.

59 Jensen, op.cit., pp. 63 ff., especially p. 73.

60 Cf. Skrubbeltrang 1938, op.cit., p. 85.

61 The principal work on the Danish crofter system is Skrubbeltrang, *Den Danske Husmand*, vol. I–II, Det Danske Forlag, Copenhagen 1952–54. His earlier work is *Husmand og Inderste*, Copenhagen 1940.

62 Skrubbeltrang, vol. I, 1952, op.cit., p. 12.

increasing number of crofters accounted for [illegible]. In [illegible] there was [illegible] crofter-class relative to [illegible] on the [illegible] while there was only a slight increase in [illegible] [illegible]

[illegible]

[illegible] sufficient [illegible]

[illegible]

Why did [illegible] place?

The [illegible] in the late 18th century [illegible]. The [illegible] was either reduced or replaced by a fixed monetary duty, and hereditary tenancy meant a limited scope for squeezing new lease-holders. [illegible] without the accompanying [illegible] property. [illegible]

[illegible]

[illegible]

[illegible] Ibid., pp. 67–68.

[illegible] Cf. [illegible]

[illegible] Cf. [illegible]

[illegible] This [illegible]

[illegible] Cf. [illegible]

[illegible] Cf. Jensen [illegible]

had already been introduced by the big landed proprietors, and freehold was attractive in itself, as free contractual tenancy on the Norwegian model never existed in Denmark. Peasant proprietorship was prepared by hereditary tenure in Jutland in the second half of the 18th century, and freehold became a political demand by coordinated peasant action in Zealand towards the middle of the 19th century, as we shall see in chapter 9.

The transition towards freehold was stimulated by the mercantilization of agriculture. Firstly because the manorial lords preferred market production and reinvestment of direct monetary profit to labour duties and incomes in kind; secondly because some freehold purchases were favoured by commercial speculation in landed property; and thirdly because the public modernization of banking and credit systems was a prerequisite for substantial parts of the purchasing process. But mercantilization was no sufficient cause. Commercial farming occurred earlier and more thoroughly through manorial farming in Zealand, which also saw the latest development towards peasant proprietorship. This was the old area of villeinage, and the sequence of change becomes incomprehensible without due consideration of the structural pattern.

Transformation of Swedish land ownership

We cannot, without reservation, specify the transition towards peasant freehold in Sweden on the basis of the questions that guided our discussion of the Norwegian and the Danish development. The Swedish movement towards peasant proprietorship took place within a different structural context, with a semi-feudal conception of land ownership, and with a classification of 'types of land' (*jordnaturer*) which did not correspond to the essential qualities of land occupancy in Norway or in Denmark (cf. chapter 5). The change of the real implication of *skattehemman* ('taxable peasant farm') was crucial in the development of freehold. The feudal notion of land possession meant a royal ultimate ownership of the peasant land. This legal conception gradually lost its practical importance throughout the 18th century, and complete peasant proprietorship, *dominium plenum,* of the *skattehamman* was codified in a particular statute of 1789.[69] The obligation to offer the farm to the Crown before it was sold out of the family no longer applied, and the royal prerogative of supervision of economic duties and farm management was dropped. Even the tenants on the Crown land, the *kronobönder,* improved their right of usage, and acquired

[69] This was done in a supplement to the new constitution of Gustav III–cf. Jutikkala, *Bonden i Finland genom tiderna,* Helsingfors 1963, pp. 318 f., and Heckscher, *Svenskt arbete och liv,* Sth. 1957, p. 186.

a formally guaranteed hereditary tenancy.[70] Thus the quality of both types of landholding was altered to the advantage of the peasantry, unlike the situation on noble lands. But the great *relative* difference between an independent *skattebonde* ('taxable peasant') and a leaseholder on Crown land prevailed.

Within this context, the transition towards peasant freehold in Sweden during the 18th and early 19th centuries was mainly expressed in the transformation of *kronohemman* into *skattehemman*, when tenants on royal estates–to secure their land occupancy–bought their holdings and changed their leasehold duties into a public tax. The first purchases of Crown land started about the year 1700, and several instances occurred relatively regularly until the right of acquisition was abolished, about the time of the constitutional upheaval, by Gustav III in 1772.[71] The purchases of tax property started anew, and continued unrestricted, when the king broke with the noble interests and drew support from the non-noble estates, including the peasantry, from 1789 (cf. chapter 5). The general development is indicated by the following figures:

The distribution of types of Swedish farms (of the total number of registered hemman *irrespective of actual subdivision). Percentages.*

Year	Crown lands	Noble lands	Peasant lands	Total
1700	35.6	32.9	31.5	100
1772	20.2	32.9	46.9	100
1815	14.5	32.9	52.6	100
1825	12.7	32.9	54.4	100

Sources: Compiled from Heckscher, *Sveriges ökonomiska historia,* vol. II, 1949, p. 271, and Heckscher 1957, p. 188.

About 2/3 of what had been Crown estates in 1700 became peasant land towards the middle of the 19th century. But the growing strength of peasant land disposal is somewhat underrated by these figures. The peasantry also acquired noble lands, *frälsejord,* throughout the 18th century–mainly scattered property outside the manorial estates. This acquisition took place partly directly from the noble owners, and partly from non-noble proprietors, who had gained noble lands earlier, and

[70] Cf. Jutikkala 1963, op.cit., p. 319. Formally, the *kronobönder* now got the *dominium utile*–cf. chapter 5 above.

[71] The abolishment of purchases was made definite in 1773, while partial suspensions had been given since 1770–cf. Heckscher 1957, op.cit., p. 188.

now found alternative investment prospects for their capital. During the last decades of the 18th century the greater proportion of noble land transferred to non-nobles was acquired by peasant farmers, who strengthened their bargaining position relative to other social classes.[illegible] The acquisition of frälsejord by peasants was legally forbidden before 1789, but the prohibition was evaded by a kind of shared proprietorship, whereby either the noble was [illegible] on a mortgage on terms that safeguarded it from redemption, or the peasant [illegible] the right of possession while the nobleman reserved the rent for himself.[illegible] The latter type of shared farmholding was called a *frälseskatte-hemman*, and the peasantry was not fully allowed to acquire noble lands until the reform of privileges in 1789. The acquisition of manorial estates and their immediate surroundings—so-called *[illegible]* (supreme noble lands)—was closed to the peasantry until 1810, when it was opened up to everybody. But even by then, about 15 % of the supreme noble lands were already owned by non-nobles, and the peasantry owned 2 % of them.

In 1845 the distribution of the total number of complete farms (*mantal*) between different ownership categories was as follows:

Ownership of Swedish farms in 1845

Nobility	11 500	17 %
Non-noble proprietors	8 000	12 %
Peasants	[illegible]	[illegible]
Church	1 000	[illegible]
Crown property	7 200	11 %
Sum	[illegible]	100 %

[illegible]

[illegible]

[illegible] peasant proprietorship in Sweden.[illegible] In 1845, the strongest

[illegible] E. Heckscher, *Sveriges ekonomiska historia från Gustav Vasa*, vol. II, Stockholm 1949, p. 27[illegible]
[illegible] see [illegible], *[illegible]*, vol. [illegible], Stockholm [illegible]
[illegible] Cf. ibid. p. [illegible]

peasant position was in the northern counties, with more than [illegible] peasant property. But in the fertile lowland areas in the southern [illegible] and [illegible], around Stockholm and on the island of [illegible], less than half of the land was peasant property. The noble landlords here [illegible] on their [illegible] [illegible] to the [illegible] [illegible] the peasants [illegible] or [illegible] of [illegible] and [illegible] [illegible] public [illegible] [illegible] In these areas of noble land [illegible] the older [illegible] [illegible] also [illegible] [illegible] and thus [illegible] [illegible] peasant freeholds. The [illegible] was also [illegible] in the [illegible] areas, where manufacturers and bourgeois landowners had been the strongest buyers of Crown lands. Land ownership was used to [illegible] from investment in timber trade and mining.

How did the 'tax purchases' from Crown lands take place? The process started on a rather moderate scale in 1701, with a royal statute that was issued to supply the means of warfare. The Crown estates were now sold as freehold land, and not as [illegible] like the [illegible] in the 17th century. The tenant on Crown land could gain [illegible] of ownership to the farm only if he could outbid the highest bidding [illegible], and thus much land was bought by manufacturers and big landowners. From [illegible], however, the [illegible] could buy their farms at a [illegible] price without having to compete at public auctions, and now peasant purchases became relatively widespread. The purchase price was regulated by a statute of 1773 to a minimum sum of six years' [illegible] rent or the third [illegible] value. [illegible] this regulation [illegible] the [illegible] since inflation made the purchases gradually more [illegible] during the 18th century. The purchase price was still modest after the new [illegible] from 1789, even if it was [illegible] to ten years' rent.

The transition towards peasant freeholds [illegible] that the peasants gradually increased their landholding status and autonomy.

[illegible]

partly by the transformation of the notion of land ownership itself, partly by the acquisition of former Crown lands, and partly by a gradual reduction of landholding barriers defined by estate privileges. The different 'types of land' tended to lose their function as separators between social estates during the 18th century. After the revision of noble privileges in 1723, higher non-noble families could possess ordinary noble lands, and from 1789 this opportunity was extended to all social estates, including the peasantry. Only the 'supreme noble lands', the manors, were reserved to the nobles, and even this privilege disappeared with the new constitution of 1810.[80]

It should also be noted that despite the tendency towards large-scale manorial farming, Sweden remained a country of peasant family farms to a greater extent than Denmark, although less so than Norway. In addition to the substantial amount of peasant freeholds, 87 % of the Crown estates and 80 % of the noble lands were farmed by peasant tenancy in 1772. The increasing subdivision of registered farms, the *hemmansklyvning,* from the middle of the 18th century, gradually favoured the transition towards peasant land control. The main exception to this picture was the highly dependent *frälsebönder* (peasants on noble land) in former Danish territories like Scania and Halland in the south. These peasants performed compulsory labour duties, *hoveri,* well into the 19th century, and thus remained marginal to the agrarian structure of Sweden, while the corresponding groups in Denmark became the subjects of an extensive reform movement as the very core of the landed nexus.[81]

Another implication of Swedish freehold is, again, the development of the crofter system and the formation of a substantial rural proletariate.

There was an increasing gap within the rural population from the middle of the 18th century, with a gradually more independent class of peasant farmers, on the one hand, and a numerous landless proletariate on the other. The farm partition did not keep up with the population growth, and the surplus population had few alternatives outside the agricultural areas. A growing number of crofter plots, *torp,* were established on peasant farms, noble estates and public common lands. The crofter, *torpare,* was a smallholder with a leasehold contractual right to his lot, in return for duties in money or kind (or both) to the landowner, and often with labour duties as well. But labour duties were generally restricted to the crofters on manorial estates, while those on

[80] Cf. Heckscher 1957, op.cit., p. 189.

[81] The structural and numerical marginality of this particularly dependent peasantry meant that their interests never became an important political issue in the Assembly of Estates. See chapters 8 and 9.

peasant land were more often spared from it.[82] The crofter system thus provided cheap labour for the landed proprietors in the south, partly as a compensation for the emancipation of the peasant farmers.

The number of crofters more than doubled during the second half of the 18th century, from less than 30 000 in 1751 to more than 60 000 in 1800, and further to a peak of about 100 000 in 1860.[83] There was also a strong growth of rural lodgers and crofters without land.[84] Both groups were highly dependent labourers and servants, but the landless crofters were also often craftsmen on their own, and paid a rent for the house. The number of landless crofters and lodgers also increased at nearly the same rate as the real crofter class during the century from 1750 to 1850, but their number in absolute terms was slightly lower.[85] The number of rural domestic servants showed a relatively moderate growth during the period, while the new group of agricultural labourers, with yearly wages and a multi-family abode, flourished with the rationalization of large-scale farming in the southern districts during the first half of the 19th century.[86]

The crofter system in Sweden, and the formation of a substantial rural proletariate relative to the landholding peasantry, can hardly be said to have been a direct consequence of the transition towards peasant freehold. The crofter system implied labour duties mainly for the 'manorial crofters', and thus became a potential substitute for–and sometimes a supplement to–the former peasant duties in the areas of large-scale farming only. But, on the other hand, the move towards freehold meant a consolidation of peasant ownership command, with a controlled limitation of farm partition. The parallel rural population

[82] Cf. Heckscher, vol. II, 1949, op.cit., p. 278. The main reference here is V. Elgeskog, *Svensk torpbebyggelse från 1500-talet till laga skiftet*, L. T.'s förlag, 1945. There were different types of crofter plots, *torp: Bondetorpare* (crofters on peasant land), *herregårdstorpare* (crofters on manorial noble lands), and *soldat-torpare* ('soldier-crofters' on peasant land, crown land, or outer commons). Some of the crofters on peasant land also had labour duties, but these were less heavy than for the crofters on the manors, and the peasant and his crofter were often also united by an original link of kinship, although status delineation increased with the intensification of agricultural production.

[83] These estimates are very approximate–cf. Carlsson in Ingers, vol. II, op.cit., p. 223, Ingers, vol. II, p. 271, and Heckscher 1957, pp. 172–173. The number of crofters declined rapidly from about 1860 onwards–to become the main group in the emigration statistics.

[84] The *backstugusittare* occupied a cottage against duties on the farm, and often managed to survive only as squatters on the outfarm commons, while the lodgers, the *inhyseshjon*, or *inderster* as they were called in Denmark and Norway, lived as lodging boarders on the farmstead.

[85] Cf. Ingers, vol. II, op.cit., pp. 276 ff., and Skappel 1922, op.cit., p. 178. In 1800 they comprised 45 000 against a good 60 000 crofters.

[86] These agricultural labourers were called *statare* or *stattorpare*. There were nearly 20 000 of them in 1833–cf. Ingers, vol. III, op.cit., p. 228.

growth then implied a clearer social [illegible] in the countryside, with substantial parts of the cottar class directly dependent upon the freeholding peasantry, even if both groups still had a privileged upper class of landed lords above them.

What was the main reason for the peasant urge to land proprietorship? The [illegible] was most obvious on the leasehold [illegible] noble lands, where the acquisition of freehold ownership meant greater security and freedom of movement. The leaseholder was [illegible] and he was spared household duties and labour services. But the situation was different on the Crown lands. The tax burden of the [illegible] and the [illegible] of his farm. But the [illegible] until the reform [illegible] when the [illegible] was [illegible]. [illegible] the extraordinary charges that were often imposed on the crown peasants, if the [illegible] meant access to [illegible] capital and a possibility [illegible] since the purchase price was far from [illegible]. Thus both the need of secure possession [illegible] and of economic mobility, represented the [illegible] of peasant freehold. The Crown sold partly because of acute financial needs and partly because compliance with the [illegible] interests constituted a balance against the [illegible] nobility that threatened the prerogatives of the Crown. Noble property and [illegible] estates could [illegible] it should be [illegible] a substantial amount of [illegible] during the 18th century as well [illegible].

The process of [illegible] the peasant [illegible] agriculture [illegible] general [illegible] in the central forest and mining [illegible] the [illegible].

Conclusion

During the [illegible] towards peasant proprietorship of land [illegible] Scandinavia [illegible] through the [illegible]

[illegible]

[illegible]

The sequences of enclosure

The Norwegian enclosure movement

There were three major degrees of communal landholding in Norway prior to the enclosure movement (cf. chapter 5). The most extreme type was complete common ownership, *sameie,* which was found even in the home fields in certain coastal areas, but which was usually confined to the 'outfields' and wooded commons.[89] Then there were often some fields with rotating allocation of use and exploitation, *årsskifte*, which was also most frequently found in the western area. Finally there was the partition of fields into intermixed strips and tiny plots, the *teig-blanding*. The allocation of strips between individual holdings was based on the possession of rent value, but the ground was usually opened up for common grazing of the cattle before the spring season and after harvest time. The proportion of communal land and the fragmentation into intermixed strips were most extreme in the western coastal area from Agder in the south and northwards right up to northern Norway. Here the scarcity of arable soil was most significant, and led to extreme subdivision of holdings within the area of the multiple farm. The boundary lines between individual farms and between separate holdings were more clearly marked in the flatland districts of eastern and central Norway, even if individual disposal was restricted by communal outfields and forest land belonging to the parish.[90] During the first half of the 19th century, the comprehensiveness of commons and intermixed strips often seems to have increased owing to farm partition and fragmentation of holdings.[91]

The Norwegian enclosures of common land and reallotment of intermixed strips most clearly belong to the second half of the 19th century, although some encloures had already taken place in the eastern lowland area during the preceding century. There was a vigorous enclosure movement from the 1860s, which culminated in the last decades of the century.[92] During the first half of the 19th century there had mainly been some enclosures of communal woods and outfields, since common ownership and rotating allocation became a hindrance to the growing

[89] Here there could be an intricate division of the right to exploit different types of resources, as we saw in chapter 5.

[90] Cf. S. Hasund, 'Grendskipnad og utskiftning', *Syn og Segn,* vol. XX, 1914, p. 290; K. Østberg. 'Fellesskap i jord og skog', *Norsk Bonderett,* vol. 11, Oslo 1936, pp. 31 ff.; Paul Borgedal, Part I and III of Torleif Grendahl, ed., *Jordskifteverket gjennom 100 år. 1859–1958,* Oslo 1959, p. 137 and pp. 360 f.

[91] A careful and detailed examination at the local level in southern Norway is presented by Hans Try, *Gardskipnad og bondenæring*. Universitetsforlaget, Oslo 1969–see p. 93 and passim.

[92] A statistical survey of reallotted area for each decade from 1860, nationally and county by conty, is presented in Grendahl, ed., 1959, op.cit., pp. 383 ff.

tendency of individual exploitation of resources outside agriculture proper. The first inquiries into the problem of communal land tenure were made by the public authorities towards the end of the 18th century. On the initiative of local officials in Norway, the bureaucracy in Copenhagen prepared several drafts of enclosure decrees around 1800.[93] This work was influenced by the contemporary reform movement in Denmark, and attention was mainly directed towards the restraining effects of communal woods and outfields in eastern Norway.[94] After the separation from Denmark in 1814, the Norwegian cabinet requested the parliamentary law committee to prepare an act on the abolition of communal land tenure. During the following years, scattered drafts on the issue were worked out by civil servants and individual farmers.[95] Finally, on 18 August 1821, the first Enclosure Act passed the Norwegian *Storting*.[96]

The Act stated that any peasant farmer could demand reallotment of intermixed strips into practicable field units, while the expenses should be shared by all partowners of a multiple farm. An assessment of real farming value was the basis of infield repartition, while corrections of the established rent value, the *skyld*, were made for enclosure divisions in the outfields. But the statute often proved to be ineffective. No competent agency was established to accomplish the task, and the peasantry were quite reluctant. The statute said that the public land tax should be increased by 1/4 if reallotment was not carried out within eight years from the coming into force of the law, but this deadline was later prolonged to 1838–and the land tax itself was abolished two years before that. The agricultural need of enclosures and reallotment seems to have been quite modest during the first half of the century, and the authorities did not do much to enforce it. In the eastern area, where the fragmentation of the home fields into intermixed strips was less extreme, some enclosures of forest land and outfield pastures were taking place.

The system of communal land tenure remained a subject of complaint–as an obstacle to agricultural entrepreneurship and innovation–from civil servants and county prefects, and a new state commission for land reforms was appointed in 1854.[97] The commission, which was dominated by local officials from western districts, gave its opinion two years later, and a new Enclosure Act was passed on 12 October 1857. Now a public enclosure service was established. The business of reallotment and enclosure was to be conducted by public com-

93 Cf. Borgedal in Grendahl, op.cit., pp. 150 ff.
94 Ibid., pp. 142 ff.
95 Ibid., pp. 154 f., and Østberg, vol. 11, 1936, op.cit., pp. 32 f.
96 Cf. Borgedal, op.cit., pp. 155 ff., and Hasund 1914, op.cit., pp. 296 f.
97 Cf. Grendahl, ed. 1959, op.cit., Part II, pp. 174 ff.

missions with a skilled foreman, if any share-owner so required. Reallotment of mixed strips was made on the basis of detailed surveying and productivity valuation, irrespective of rent value – while outfield commons were enclosed according to the official rent assessment of 1821, provided no other ownership relations were established.[illegible] The local enclosure commission could also instruct one or more peasant farmers to move their houses out of the multiple farmstead nucleus, if this was found necessary to achieve a more [illegible] concentration of the individual holdings. The expenses were to be shared among all part-owners relative to the rent value of their farm, and the process was subsidized by the state.[illegible] The Act of 1857 thus facilitated the enclosure movement which accelerated all over the country from the 1860s onwards.

Norwegian land reforms had by now become [illegible] urgent. The [illegible] and led to increasing subdivision of farms with [illegible] fragmentation of strips and plots. The need for more arable land meant increasing incorporation of the outer commons into the home field area, and enclosure was the only feasible solution. More efficient exploitation of the arable land implied new farming methods and tools. From the 1850s and 1860s, crop rotation and alternation between meadows and tilled fields increasingly replaced the old open-field system (cf. chapter 7 below). The introduction of the iron plough gradually made the tilling of meadows possible, and reallotment became a general demand.[illegible] The [illegible] among the peasantry does not seem to have [illegible] and [illegible], but was rather directed against the compulsory removal of houses from the multiple farm nucleus. This meant a radical dissolution of the close community relationship, but was often a necessary price to be paid for that greater management [illegible] which the [illegible] agriculture required. At this stage the peasant [illegible] did not [illegible] question the principle of [illegible] should be [illegible]. In the next chapter, the agricultural revolution in Norway and the discussion of its relation to the transition to [illegible] will be [illegible].

[illegible]

[illegible]

[illegible]

[illegible]

The first complete enclosures and redistribution of the [illegible] village lands in the kingdom of Denmark took place in the 1760s. [illegible] manorial estates had in many areas been separated from the peasant village lands already in the 16th century. South Jutland and the Duchy of Schleswig are the main exceptions, where [illegible] and [illegible] through the [illegible] 17th centuries. [illegible] is supposed to have been affected by the early extensive [illegible] of large-scale farming, where the [illegible] enclosed land.[illegible] The first separation of [illegible] village in South Jutland took place in 1608, and [illegible] during the first decades of the 18th century.[illegible] In most parts of the kingdom, the enclosure of peasant farm land did not start until the 1770s, and the whole process seems to have culminated around the turn of the century.[illegible] [illegible]

The Danish [illegible] movement [illegible] land-owning [illegible] saw the reform as a prerequisite for [illegible] progress in the countryside. By the middle of the 18th century, the government [illegible] a number of [illegible], and the first public [illegible] were [illegible] 1757.[illegible] The big [illegible] resulted [illegible] forming of [illegible] estates, and the enclosure works then [illegible] increased the labour duties of the peasantry.[illegible] During the next thirty

[illegible] Cf. Troels-Fink, [illegible], Munksgaard, Copenhagen [illegible].

[illegible] South Jutland, with the early enclosure movement, [illegible] the peasantry [illegible] agriculture [illegible].

[illegible] Ibid., pp. 83 ff.

[illegible] Cf. C. Rise Hansen and Axel Steensberg, [illegible] *Undersøgelser [illegible]*, Munksgaard, Copenhagen [illegible] detailed investigations in the county of [illegible] presented in *Historisk Samfund for [illegible]* [illegible].

[illegible] Skrubbeltrang 1938, op. cit., pp. 61–62.

[illegible] Ibid.

[illegible] Cf. Rise Hansen and Steensberg 1951, op. cit., p. 352. This was the outcome of the first public land commission [illegible].

[illegible] Cf. Skrubbeltrang 1938, op. cit., p. 45.

years, gradually more radical statutes were issued.[109] The first ones stated that the principal owner of the village *hartkorn* could demand repartition and enclosure towards neighbouring villages, and that any part-owner could enclose his share of the outfields. A statute of 1769 then declared that any part-owner could carry out repartitioning and enclosure of his own fields and meadows at his own expense, while all the part-owners should share the costs of outfield reallotment. Then in 1776 it was legally provided that any part-owner could demand that his fields be repartitioned into maximally three consolidated shares through the mediation of an authorized land surveyor and a land inspector, and that the costs should be shared by all owners on the basis of a public valuation. The most important of these statutes was issued in 1781. Any village part-owner had a right to demand that his land be repartitioned into maximally 2–3 coherent shares, and a project of complete village reallotment should be worked out even if only one estate owner demanded repartition of his peasant farms. The expenses were to be shared by all part-owners, including those who preferred to remain within common village lands. Yet the business of enclosure and reallotment was a rather expensive affair, and there was still some reluctance among the landowners.[110] This reluctance was finally overcome by economic incentives through a statute of 1792, after which village part-owners who repartitioned and consolidated their estates could claim an interest of 4 % of the original capital cost from leaseholders and crofters whose lands were included in the reform. They were also allowed to increase the rent, the *landgilde*, after reallotment of peasant farm land. These measures particularly seem to have made the enclosure movement go ahead.[111]

It should be noted that the Danish enclosures, and reallotment of intermixed strips, were part of the general agrarian reforms of the last decades of the 18th century. Peasant farming was so impoverished by heavy duties and obligations that it became an economic strain even for the landowning class. The break-up of the open-field system–and the subsequent change from two- or three-field farming to crop rotation–was connected to the 'peasant emancipation' from above–the development of copyhold tenure and the dissolution of compulsive residence and villeinage. The succession of reform actions was not without consequence. The prospects of progress were probably greater where the land of some peasant farms was consolidated by the landlord into coherent shares before the labour duties were replaced by a fixed monetary rent and leasehold transformed into copyhold tenure, than if

109 See Rise Hansen and Steensberg 1951, op.cit., pp. 352–356.

110 Reallotment and removal were to some extent, however, supported by reasonable loans from the Credit Bank of 1786, and encouraged by prizes from the agricultural society, the *Landhusholdningsselskab*.

111 Cf. Rise Hansen and Steensberg 1951, op.cit., pp. 255–256.

these reliefs were given to poor peasants on badly farmed commons.[112] From the 1780s, this body of reforms was accomplished by landlord pioneers as well as on the remaining Crown lands.[113]

The enclosure movement entailed some removal of individual farms away from the village nucleus. This was partially counteracted by the use of stellate reallotment as a common repartition figure, whereby the boundaries of each holding radiated from the village cluster. But where stellate reallotment was not adequate or feasible, as was often the case, repartition meant consolidation of fields into more or less rectangular blocks–and the village community ceased to exist as a natural ecological centre.[114] By 1802 it is estimated that nearly 40 % of the peasant farms in Zealand had moved out of their former village clusters. This was a substantially greater proportion than the tendency in other parts of the country, but in Jutland the single farmsteads were more numerous beforehand.[115]

The land tenure reforms did not imply an unstinted and egalitarian growth of potential peasant power. For some parts of the peasantry it obviously did, but for others it did not. The differentiation of the peasant community increased. Small-holding peasants were often offered only outfield lots of the village lands, and they lost their participation in the community fellowship together with the rights of exploitation of the commons. These effects also hit the crofter class, who became substantially more numerous during the reform period. Provided they carried out reallotment and relieved the peasants of *hoveri* (labour duties), the landlords were often licenced to close down peasant farms and rejoin them to the manorial estates. The lands were then parcelled out again into smaller crofter plots to replace the lost labour duties of the peasant farmers, since the crofters were subject to more arbitrary domain.[116] The public authorities made some effort to ensure that the crofter class received plots of land as a compensation for the lost access to the commons, but the crofters were on the whole abandoned to the control of the landowners, and the class division within the peasantry generally increased.[117]

112 Cf. Skrubbeltrang 1938, op.cit., p. 46 and pp. 63 ff.

113 Ibid., pp. 50–51.

114 Cf. Rise Hansen and Steensberg 1951, pp. 365 ff. and p. 388.

115 Cf. Skrubbeltrang 1938, op.cit., p. 62. The more vigorous removal from the village clusters in Zealand might be due to the stronger landlord dominance in this area. It was particularly the biggest villages that were dissolved.

116 Cf. C. Christensen, *Agrarhistoriske studier II. Danske landboeforhold under enevælden*, I, Copenhagen 1891, pp. 13–14. Christensen states that in Zealand in the year 1800 peasant farms were parcelled out into about 30 000 new crofter plots of the size of from 1 to 10 'barrels of land' (*Tdr. land*).

117 Ibid., p. 17; cf. Skrubbeltrang 1938, op.cit., pp. 46–48. Skrubbeltrang seems to imply that the more advanced the enclosure movement, the more numerous was the crofter class–ibid., pp. 67–68.

Since the later decades of the 19th century, Swedish agriculture has been characterized by dispersed settlement, with isolated farmsteads or very small hamlets. Village communities have persisted only in a few provinces like Dalarna, but as we saw on p. [illegible], there were villages in most of Sweden around the 18th century. Each village consisted from about 2 to 20–30 individual farms, and [illegible] were most extensive on the [illegible] plain land, while the village clusters were more common [illegible] the [illegible] against [illegible]. While the [illegible] and the outfields were often common property, arable and meadows were held in long, narrow and regular strips of land, which were separated by a [illegible] or an uncultivated strip. The holdings of each farm consisted of numerous strips, [illegible] parts of [illegible]. The [illegible] of village [illegible] area of common land, [illegible] of [illegible] strips and [illegible] into larger pieces for each farm, and dispersion of the village [illegible] started around the middle of the 18th century.[illegible] Agronomists, land surveyors and other [illegible] the village community and its old tenure system as an obstacle to efficient agriculture, and the authorities were persuaded to initiate land reforms.[illegible]

The first stage in the Swedish land reforms, the so-called *storskifte* (meaning 'consolidation into large parcels'), was based on a statute of 1757.[illegible] Eight years earlier, it had been legally prescribed that [illegible] of mixed strips into larger pieces could take place provided all part owners agreed, but now [illegible] and from 1762 every peasant farmer could demand *storskifte* to be arranged in his village. The reallotment was conducted by a public land surveyor, and the plan was to be ratified by the local courts.[illegible] The less radical alternative was a consolidation of one or more farms into a few single

[illegible]

[illegible] The following accounts have [illegible] from Heckscher, vol. [illegible] and Heckscher [illegible] Wohlin 1912 [illegible] vol. [illegible] and vol. III [illegible] Montgomery [illegible] 1933.

[illegible] of the Land Survey office, Jacob Faggot, [illegible]. Faggot's main objective was to replace the multiplicity of small strips with large, consolidated holdings. As early as [illegible] p. 270.

[illegible] 1965, p. 259. [illegible] p. 60.

pieces, while the rest of the village either kept its old strip and field system, or just reduced the number of strips. Ordinarily, there was no depletion of the nucleated villages during this early period of repartition and enclosure.[123]

The *storskifte* usually resulted in a reduction of the number of strips, and a concentration into bigger plots, although complete consolidation of individual holdings seldom occurred. Only the manorial estates preferred a concentration of all the land into one single piece, and made their strips into coherent areas. This practice was legalized in a statute of 1783.[124] The famous master of the large manor of Svaneholm in Scania, Baron Rutger Maclean, extended this practice even to his peasant leaseholders, and carried through complete redistribution of the holdings of his ca. forty peasant farmers in the middle of the 1780s. The four villages within the boundary of the estate were all broken up, and the farm land was concentrated around new scattered farmsteads.[125] This radical measure was originally opposed by the peasantry, but became the prelude to the second stage in the Swedish enclosure movement–the *enskifte*. This implied a repartition of the soil into one coherent share for each farmer (which had also been the original intention of the advocates of the *storskifte*), and entailed moving the houses of each farm from the village cluster to new sites apart from the others. The *enskifte* was first initiated on the plains of Scania and Västergötland, where it was regulated by a statute of 1803. Four years later the practice was given legal sanction in all southern and central provinces.[126]

However, the real solutions were often somewhat less radical than those provided by law, and the authoritarian methods of Baron Maclean and other nobles were unsuitable in regions of peasant freehold. The *enskifte* was adapted to flat country areas, but too extreme in heterogeneous terrain with wooded plots and sparse arable land in narrow valleys or broken country.[127] The experience of Swedish enclosures thus disclosed that while the *storskifte* had been too ineffective, the *enskifte* was too inflexible. The outcome was the statute of *laga skifte*, 'legal repartitioning', of 1827, which then marks the third stage in the Swedish land reforms.[128] It was no longer obligatory for all farms to move out of the village cluster in the case of repartitioning, and outside its main consolidated piece, a farm might get additional pieces of land apart from each other.

[123] Cf. Jutikkala 1963, op.cit., p. 272.

[124] Cf. Dahl, *SEHR* 1961, op.cit., p. 62.

[125] See Ingers, vol. II, chapter XIII, pp. 435–460.

[126] Cf. Ingers, vol. II, ch. XIV, pp. 461–503; Heckscher 1957, op.cit., pp. 176–177. Dalarna and Norrland remained exempted–cf. Wohlin 1912, op.cit., p. 597.

[127] Cf. Dahl, *SEHR* 1961, op.cit., p. 64, and Heckscher 1957, op.cit., p. 177.

[128] Cf. Albert Eskerød in Ingers, vol. III, op.cit., pp. 19 ff.

It has been estimated that the *storskifte* during the second half of the 18th century about halved the extension of mixed fields and commons.[129] The land reform was on whole accomplished by 1870. The village communities were broken up through the *enskifte* and the *laga skifte* during the first half of the 19th century.[130] But there were great regional and structural variations. The initial enclosures, and the most complete ones, took place on the southern plains with clear landlord dominance, while the traditional strongholds of peasant freehold–in areas of wooded and broken country further north–showed a slower and more irregular transitional sequence. The complicated villages of Dalecarlia were even exempted from the *laga skifte*, and have kept the *storskifte* pattern right up to the present day.[131] The Swedish land reforms were never officially mandatory if no peasant farmer supported them, but they were strongly encouraged by the authorities, and they were initiated by great landowners and rural lords. The manorial estates were particularly hampered by the dependence on the village neighbours, and the state provided the technical assistance in land reallotment. The reforms were only later accepted by the peasant farmers, whether they were freeholders or tenants.

The Swedish land reforms were primarily stimulated by a dual advantage–through processes of an economic and demographical nature.[132] The repartition of mixed strips and enclosures of common land meant a more free scope for individual enterprise. Agricultural innovation became independent of the farming practice of the neighbours, and new soil could be cultivated when individual ownership of the outfields was established.

In an agricultural development that was gradually more geared towards production for the market–with a reinvestment of surplus partly in agricultural innovation and partly in trade and industry outside agriculture–the village system with its mixed strips and fields complicated the economic mobilization of the soil. The empirical accuracy of this interpretation–which implies that the stronger the subsistence orientation of the peasantry, and thus the weaker the entrepreneurial component of farming, the weaker the urge towards reallotment and enclosure–will be further illuminated in the next chapter. At this point it can also be affirmed that the enclosures of wooded land stimulated forestry and lumbering–and vice versa.[133]

129 Cf. Jutikkala 1963, op.cit., p. 275. This is obviously a very approximate estimate, and exact calculations are impossible to make.

130 Cf. Wohlin 1912, op.cit., p. 601. Wohlin gives the proportion of total land area that underwent reallotment, but this says nearly nothing about how throroughly the *arable* lands were transformed–ibid., pp. 599–600.

131 Cf. Dahl, *SEHR* 1961, p. 67.

132 Cf. Wohlin 1912, op.cit., pp. 602–605, and Montgomery 1931, op.cit., p. 35 and p. 43.

133 Cf. Montgomery, op.cit., pp. 51–52.

The dissolution of narrow intermixed strips also had some immediate economic advantages. The old system hampered effective draining and sawing by new machinery, and the reforms involved abandonment of the old two- or three-field system, since each farmer now became free to use any crop rotation he pleased. The area of arable land was directly increased by the removal of demarcation dikes and ditches, and the waste of labour time through constant change between distant strips was eliminated (cf. chapter 7 below).

But the land reforms were also stimulated by the development within the peasantry itself–i.e. the population growth and the problems of inheritance in a society with few alternatives outside agriculture. In many instances the *storskifte* was a prerequisite for further farm division, *hemmansklyvning*. The narrow strips and scattered tiny plots were often unsuitable for further splitting, and the *hemmansklyvning* was often made economically practicable only through the rationalization of farming which was inherent in the land tenure reforms. The reallotment and enclosures in Sweden were thus often blended together with farm partition and division of the estate.[134]

But the land reforms also implied obvious disadvantages–generally to the peasantry as a whole, and particularly to the poorer sections of the peasantry. The reforms destroyed the close communitarian fellowship and the collective security of the fused village communities. Mutual help and common labour had to be replaced by individual entrepreneurship. The common ownership and intermixed land tenure had also been a guarantee against conspicuous mismanagement, and was not only an obstacle to initiative. The protracted business of complicated repartition, during which obligations and ownership relations were diffuse, also resulted in some neglect of proper farming.[135]

Some specific and inevitable reform problems also provoked peasant resistance. Since the different fields were of different quality, it was difficult to compensate a clear loss of good soil with greater quantities of less valuable land. The public land surveyors here generally used the land rent assessment, the *øre-tal*, as the criterion of repartition, but this was a poor help in the case of newly cultivated fields, where the fixed rent assessment was highly disproportionate to the real land value.[136] There were also bitter conflicts about which families should first move out of the village nucleus, and how the new settler buildings should be paid for.[137]

134 Cf. Heckscher 1957, op.cit., p. 179, and Wohlin 1912, op.cit., pp. 602–603.

135 Jutikkala 1963, op.cit., p. 271.

136 Ibid.

137 The costs of land repartition meant that the accomplishment was influenced by economic cycles–stimulated by good times and hampered during agrarian depressions–cf. Utterström, *Jordbrukets arbetare*, vol. I, 1957, pp. 521–522.

These problems constitute the background of the particularly efficient resistance—and the relatively late accomplishment of the land reforms—in areas with many part-owners and a relatively even distribution of land. Here no dominant landlords were present to enforce repartition against a reluctant or openly resistant peasantry. Other transforming forces had to make their way within the peasant community itself.

It should also be stressed that the small peasants and the outfield squatters, the *backstugusittare*, were particularly victimized by the enclosures. These groups had formerly been free to exploit the outfield resources, but were now assigned to negligible shares according to their legal ownership rights. They also had to shift for themselves, in isolation from the former fellowship.[138] Further social differentiation of the peasantry was thus also a consequence of the land tenure reforms. As in Denmark, they seem to have stimulated the development of the crofter system.

Recapitulation

The Scandinavian land reforms were a prerequisite for technical innovation in farm management. Draining, crop rotation, and the efficient use of mechanical tools required bigger consolidated fields, and individual enclosure meant greater independence from more indolent and sluggish neighbours. The old communitarian fellowship was based on strongly egalitarian norms, and did little to stimulate individual initiative. The land reforms paved the way for expansive agriculture, producing for the market.

The Norwegian reforms started relatively late—nearly a century behind those of Denmark and Sweden. But the Danish enclosure movement, and the initial Swedish one, were led by the manorial estates, which were practically unknown to the Norwegian agrarian structure. In Denmark the enclosures occurred as part of the general 'peasant emancipation', which was strongly motivated by the public need of increased tax incomes. It could also be argued that the Norwegian reforms could be postponed because of ecological conditions. The strips and plots of different owners were usually scattered between meadows and pasture land, since arable soil was scarce. They could often be farmed without having to traverse the neighbour's land, and thus the strip mixing was less dense than was the case in the rigid three-field system of flatland Denmark and southern Sweden. The strict village regulations of Denmark and Sweden, with compulsory management systems, the *driftstvang*, were never necessary on Norwegian lands.

The social implications of reallotment and enclosure also seem to be quite dissimilar. The land reforms of Sweden and Denmark stimulated

138 Cf. Wohlin 1912, p. 606, and Heckscher 1957, p. 180.

the formation of a substantial rural proletariate through the crofter system–dispossessed squatters and poor peasants who had lost their former access to the village commons and outer woods and hayfields. In Denmark the crofter class even became a substitute for the formally liberated peasantry–as a cheap labour force on the manorial estates–through the same reform movement that accomplished the enclosure of communal land. The enclosure movement in Norway, on the contrary, coincided with the gradual dissolution of the crofter system. The economic transformation which gradually made agriculture into an individual enterprise, also released the rural crofters from non-economic dependence, and attached them to the labour market as worker-peasants or rural migrants.

Throughout Scandinavia the enclosure movement proceeded by the direct intervention of the state. Agricultural 'modernization' became a central engagement in public policy–in accordance with physiocratic theory–although the landed interests influenced the reform procedure in different ways. While landlordism pushed for a vigorous process in Denmark and southern Sweden, the Norwegian officials had to wait for the time when the reforms were urged by the transitional peasant economy itself.

The individual family farm became the basic unit in the agricultural pattern all over Scandinavia, in sharp contrast to the British enclosures, which rather stimulated the final elimination of peasantry as a social formation. Scandinavian enclosure occurred at a time when there was no substantial outlet for the population in industrial work–as had been the case in Britain–and it furthered farm partition and increased peasant farming rather than the abolition of peasant land. Large-scale farming on the English model had never become so dominant in Scandinavia, and enclosure therefore had very dissimilar effects. There was no mass of dispossessed village labourers to be drained off the land, but rather a substantial number of peasant leaseholders who acquired individualized freehold. But even in Scandinavia, the peasantry gradually lost some of its originally intrinsic characteristics, since it came to constitute a professional occupation more than a versatile social group, an entrepreneurial 'farmer' type more than a subsistence-oriented 'peasantry'. The overall structural pattern becomes, however, incomprehensible without accounting for the regional contrasts in the range and sequence and character of the process.

Peasant proprietorship and individual land tenure

We have given a comparative survey of the transition towards 'possessive individualism' in land, as it manifests itself in peasant freehold and enclosure. The diversified character of the process has been demonstrated, although there are also definite similarities in the pattern of events.

It is difficult to establish any exact dating of the developmental relations between peasant proprietorship and land reallotment. Partly because the processes occurred gradually, with very indistinct points to mark the beginning and the end, and partly because there were extreme regional variations.

In Norway the main move towards freehold took place prior to the enclosure movement. The transition towards the new peasant proprietorship started very early, in the last decades of the 17th century, but lasted for a long time–until the end of the 19th century in some regional areas. Enclosure, on the other hand, belongs principally to the decades from about 1860, and was accomplished on peasant freeholds.

The transitional sequence in Denmark was nearly the opposite. Peasant proprietorship came relatively late, cautiously from about 1760 onwards, but was carried out before the middle of the 19th century. Enclosure of peasant farm land was part of the reform movement during the last decades of the 18th century and the first decades of the 19th. Freehold was acquired on the basis of consolidated pieces of individual peasant farms.

The Swedish processes occurred more or less parallel to each other, and might somehow be dated between the sequential pattern of Denmark and Norway. The move towards peasant freehold (which did not yet imply full proprietary rights) started in the beginning of the 18th century, and was accomplished some few decades into the 19th century, although a considerable number of noblemen's estates still continued to exist. Enclosure started, at least partially, around 1750, and lasted for about a century.

But this rough comparison on the level of the nation-state is quite misleading if it is not broken down regionally. Then the presence or absence of landlordism seems to have been decisive. The initial Swedish enclosure movement took place in the southern area with large-scale manorial farming, while village commons and intermixed strip farming prevailed considerably longer in the traditional strongholds of peasant freehold. Norwegian versus Danish agriculture fits in to this pattern. Danish landlordism implied individual entrepreneurship *per se*, and stimulated enclosure of peasant land according to its own economic demands; while the Norwegian enclosure movement–accomplished by a peasantry without rural lords above them–required an internal transformation of peasant agriculture as such. This will be further specified in the next chapter.

The crofter system was a transitional phenomenon. It grew out of the socially discriminating mechanism which was implicit in the formal establishment of new proprietary rights. The conditional background was the rural population growth and the labour-intensive mode of agricultural production. The Norwegian crofter class was tied to the free peasant proprietors, while most crofters in Denmark and southern Sweden belonged to large-scale manorial farming–as a substitute for peasant obligations. As demonstrated in western Norway, the repressiveness and social inequality of the system was subdued in the case of a generally closer fellowship in the peasant communities.

CHAPTER 7

The agricultural revolution

In the previous chapter, we suggested a close connection between technical factors and methods of farming on the one hand and the reallotment of mixed strips and commons on the other. The processes of change which affected farm management will be outlined in the present chapter.

Some Scandinavian areas, especially in Denmark and southern Sweden, had already experienced a notable agricultural revolution during the second half of the 18th century, but generally the growth of neotechnical agriculture, together with the expansion of the market, took place in the 19th century. The structural pattern varied greatly according to regional ecological factors, but agriculture on the whole changed its basic characteristics, and the peasantry became a new type of societal actor.

The alteration of tools, equipment and farm management was a prerequisite for commercial agriculture and the intrusion of the market. New demands for credit supply and commodity distribution systems evolved as the market expanded. The peasant society slowly adapted itself to the new economic conditions by special associational devices. and the development also had definite consequences for the rural social structure.

Even during the 18th century and earlier, the peasantry had a certain need for cash–for some of the taxes and duties, and for additional requirements like iron, salt, and special equipment. The market sale was often concentrated on the outcome of subsidiary occupations like fishing, forestry or handicrafts, especially in Norway and substantial parts of Sweden. But even grain and livestock were bartered between surplus and deficit areas. Throughout the 19th century, the economic significance of secondary occupations was relatively reduced in the best agricultural areas, while they came to constitute the main means of livelihood in agriculturally marginal districts. Functional specialization of trade and crafts is an important aspect of the market integration of the peasantry.

The substantial population growth, which was documented in chapter 5, stimulated an extensive land reclamation during the first decades of the 19th century, but later it had to be answered by more intensive rationalization of agricultural èquipment and farm management. A fragmentation of holdings occurred all over the Scandinavian area, although the vast emigration to North America came to constitute a safety valve during the last decades of the century.

This chapter is drafted on the opinion that a direct quantitative comparison of the developmental pattern for technical innovations would be somewhat misleading. The diffusion of a new plough does not–owing to different structural and ecological conditions–have the same implications within each national setting. In addition, the available sources do not have identical qualitative characteristics, and some of the primary research on which the following account is based, have proceeded from different methodological angles. A strict comparison of quantitative indices would falsely obscure these shortcomings.[1]

The new farming

Norway. Although the most notable transformation of Norwegian agriculture took place during the second half of the 19th century, there had been a moderate technological progress, with investments in machinery in most parts of the country, during the first half of the century as well.[2] Norwegian agriculture probably was, on the whole, technically backward by North European standards around the year 1800. It was primitive even by comparison with most of Sweden, and even more so compared to Denmark.

Two major types of agricultural methods dominated Norwegian lands prior to the slow intrusion of neo-technical farming during the 19th century.[3]

[1] One of the most outstanding Norwegian studies of agricultural innovation during the first half of the 19th century is Fartein Valen-Sendstad, *Norske landbruksredskaper, 1800–1850-årene,* Lillehammer 1964. This work is extensively based on the records of partitions of inheritance, and has no real counterpart in Swedish or Danish historiography. The various studies use different types of primary sources (material remnants, contemporary accounts, official records etc.)–which means that comparative figures and information should be treated with the utmost care.

[2] Agrarian historians like Simen Skappel, Sigvald Hasund and, most explicitly, F. Valen-Sendstad (op.cit.) have noticed the gradual technical transformation which took place from the very beginning of the 19th century. This view is opposed to those theories of standstill which probably dominated the historical view until quite recently, and which dated the transitional stage more or less exclusively to the second half of the century–cf. in particular W. Keilhau, *NFLH,* vols. VIII–IX, Oslo 1929–31; O. Klokk, *Det norske landbruk siden omkring 1750*, Kra. 1920; Jonas Smitt, *Det norske Landbrugs Historie*, Kra. 1875. Valen-Sendstad pushes this debate to extremes in his introductory chapter, op.cit., pp. 8 ff.

[3] Cf. K. Visted and H. Stigum, *Vår gamle bondekultur,* vol. I, Cappelen, Oslo 1951, pp. 146–148, and Valen-Sendstad 1964, op.cit., ch. I. The local variations and transitional forms were quite numerous within these major types.

The *one-field system,* called *reitbruk* or *ensædesbruk,* implied that the fields were heavily manured each year, and continuously cultivated without fallow years. This method required a substantial livestock relative to the tilled fields, since artificial fertilizer was not available. There existed an optimal equilibrium between field width and livestock, and there was a permanent separation between fields and meadows. The small field strips of this system were thoroughly tilled by hand tools, and gave relatively good yields. The one-field system was prevalent in the south-western area, but with some modifications it also dominated the communities both farther south and farther north along the western coast. Within the most typical area of the one-field system, in the south-west, the soil was tilled by spade, hand rake and sledge. Certain intermediate areas northwards and in the valleys farther south and east showed some early instances of ploughing, but the ploughs were wholly wooden, with straight mould boards. The one-field system was, generally, confined to areas with relatively egalitarian farm units, and with a system of strongly intermixed field strips which acted as a barrier towards technical innovation.

The second major type was a *multiple-field system* with more or less regular fallow years. A four-field system was quite usual, with one fallow year rotating between four parts of a cultivated field area. Some areas also had an ordinary continental three-field system (like the German *Dreifelderwirtschaft* which was widespread also in Denmark and southern Sweden), but perhaps more often there were five or six fields with one part yearly in a rotating fallow. A more irregular rotation method, which was also employed in some eastern districts, was continuous cultivation of the same seed until the field became useless, whereafter it was turned into fallow for a longer period, to recover from exhaustion. This method was sometimes used for oats (and therefore called *havreskifter*) parallel to more regular rotation systems.

The multiple-field system was employed in the plain area in south-eastern and central parts of the country, where corn production was agriculturally dominant. The fields were sparsely manured, and thus required a smaller amount of livestock relative to the tilled field area. These lowland fields were bigger and more consolidated than the scattered intermixed strips in the west and south-west, even if yields per acre were substantially poorer.[4]

In the eastern and central area, tilling by spade was employed on some crofter plots and horseless smallholdings only. Farming was ordinarily done by ploughs with wrought iron mould boards, supplemented by harrows instead of rakes, and wheeled equipment (carts and wagons) instead of sledges. These areas had better external communica-

[4] Cf. Valen-Sendstad, op.cit., ch. III.

tions; they traditionally kept more horses because of the forestry; and they showed a stronger concentration on agriculture proper–where the labour force was less diverted towards fishing and other non-agricultural means of livelihood. The bigger farms close to the cities (especially outside the capital), as well as other 'model farms' in the east, used substantial manuring, fodder production and cultivation of the meadows already at the dawn of the 19th century. There were on the whole more varied conditions, with more differentiated agricultural standards, in this area. The holdings were less egalitarian, and cultivation methods less static, than was the case in the one-field spade area further west. Eastern agriculture seems more dominated by market integration, which particularly favoured the hinterland of the urban centres. The outer communications were also generally better compared to the fjord and mountain area.

In the south-east and Trøndelag there had already been a slow agricultural transformation during the second half of the 18th century. This was probably stimulated by the relative lack of intermixed strips and by the early enclosures of the commons, as well as by the early transition towards peasant proprietorship, which particularly comprised the east. Some impulses might also have diffused from the scattered model farms of officials and bourgeois proprietors, although their innovations were principally based on reinvestment of surplus gained in lumbering and other trades rather than from agriculture itself. Some impulses, however–like iron ploughs and harrows–met functional requirements which had evolved within peasant agriculture itself.

New soil was intensively cultivated during the first half of the 19th century, to meet the population pressure. Although land reclamation might have been the most notable method of agricultural amendment, there was also a gradual improvement of farming methods, which led to a rapid increase in output.[5] In the east, the centre (Trøndelag), and the south-west (Jæren), the investments in new agricultural equipment aimed at increased efforts in agricultural production as such, while the motives in other areas–especially along the coast northwards–must rather have been to release labour force for other tasks.[6] Widespread land improvement in the form of field draining and stone clearing took place in many parts of the country from about the second quarter of the 19th century.[7] The agricultural elite among the bourgeois landed proprietors imported new tools and methods from abroad, and be-

[5] Cf. Steen 1957, op.cit., pp. 24 ff. See also F. Sejersted, *En teori om den økonomiske utvikling i Norge i det 19. århundre*, Stencilled, Oslo 1973, pp. 29 ff.

[6] See Valen-Sendstad, op.cit., pp. 204 f. and passim.

[7] The typical stone walls of Jæren, and the stone stables of the eastern valleys, are visible relics from this strenuous movement.

came an officially encouraged spearhead of the innovative activity. Threshers became quite widespread in the eastern plain area from the 1820s, and grain drills also conquered eastern districts with a favourable market location and good communications. The same applied to the new light swing ploughs and the rolling iron harrows, as well as to the mowers and reapers which began to displace the scythe in the east from the 1850s.[8] It should be noted here that the eastern urban areas were supplied with iron manufacture and a standardized series production of agricultural implements for an anonymous market as early as the beginning of the 19th century. The breakthrough for an agricultural machine manufacture–producing articles like threshers–seems to have taken place in the relatively advanced south-east around the middle of the century, while the peasantry in the rest of the country –and particularly in the subsistence-oriented periphery–still produced their own tools on the farm, or gradually developed more specialized rural handicraft.[9]

During the 19th century, the plough gradually conquered the 'spade area' of the south-west. The wooden plough with the straight mould board, well known in some western districts further north, often came first. Around the middle of the century it was replaced by the wrought-iron plough. This transformation of the tool-culture at first affected those bigger farms that were favoured by the market–as was generally the case in the transitional pattern for the country as a whole.[10] The sledge was also replaced by farm carts and wagons. In the eastern districts, this wheeled equipment had early been stimulated in agriculture, since all long-range transport had traditionally taken place by land, while the western communications at sea meant no such stimulus. The innovations of agricultural tools and equipment in the west were also relatively retarded owing to extensive intermixed field splitting and tenancy conditions.[11] The south-western district of Jæren absorbed equipment for horse-power with the greatest force towards the middle of the century, since competition from fishing was negligible here compared to the coastal area northwards, and since market conditions became increasingly favourable with the growth of the city of Stavanger –whose population doubled twice from 1801 to 1855.[12]

Despite the reform of agricultural practice that had taken place in the country as a whole, harvesting methods were still backward. The

[8] Valen-Sendstad, op.cit., pp. 176 f. and pp. 198 ff. The official registration counted in all 1299 mowers and reapers in 1875–cf. *Historical Statistics 1968*, Oslo, table 96, p. 147.

[9] Valen-Sendstad, op.cit., ch. V, pp. 235 ff.

[10] Ibid., ch. III, pp. 120 ff.

[11] Cf. ch. 6 above. See also Valen-Sendstad 1964, pp. 208 ff.; Paul Borgedal, *Norges Jordbruk i nyere tid*, vol. I, Oslo 1966, p. 76; Steen 1957, op.cit., pp. 35–43.

[12] Valen-Sendstad, op.cit. pp. 227 f.

grain was generally cut by sickle during the first half of the 19th century, and only some eastern districts changed to the scythe from the 1830s.[13] The real breakthrough for mowers and reapers, grain drills, steam-threshers, and wheeled horse rakes had to wait until the last quarter of the century.[14]

During the 19th century, the old agricultural methods–the one-field *reitbruk* and the multiple-field fallow system (*tredebruk*)–were gradually replaced by ordered rotation of crops. This implied greater ploughing depth and a regular cultivation of meadows. The new agricultural implements were thus a basic technical prerequisite.[15] The first tendency was an increasing rotation between grain and potatoes. A real crop rotation between corn, roots (especially turnip) and grasses (clover and timothy) developed later. This was accompanied by better fertilizing, draining and tilling of meadow land. Better cereals, like wheat and rye, also became more widespread at the expense of oats, and it has been estimated that yields on the average nearly doubled relative to sown seed during the first half of the 19th century.[16] There were many transitional types between the older methods and the systematic crop rotation–often lumped together under the notation *kobbelbruk*–which prevailed in many peripheral districts towards the dawn of the 20th century.[17]

To sum up, it thus seems that agricultural methods were more technically backward if the holdings were small, fragmented by intermixed field strips and commons, and farmed on leasehold contracts; and if agriculture was subsidiary to other occupations (thus mainly left to women and old folks), with unfavourable market conditions. The iron plough and the wheel, those symbols of the Norwegian 19th century neo-technical transformation, spread gradually from the bigger, consolidated freeholds, favoured by near-by markets and without serious competition from subsidiary occupations. The innovations were internal to that structural change which gave a greater part of Norwegian agriculture these characteristics.

Around the middle of the 19th century there also occurred a notable improvement in the management of livestock, when breeding and

[13] Borgedal I, op.cit., pp. 138–139. Borgedal was a professor of agronomy at the Norwegian Agricultural College. His historical work (op.cit.), in three volumes, is critically reviewed by Kåre Lunden, *Historisk Tidsskrift*; vol. 49, no. 1, 1970, pp. 91–98.

[14] Borgedal I, op.cit., pp. 156–157.

[15] Valen-Sendstad, op.cit., ch. III, passim.

[16] Cf. S. Hasund, 'Korndyrking i Noreg i eldre tid', *Bidrag til Bondesamfundets historie*, I, Oslo 1933, p. 201.

[17] Cf. Hans Try 1969, op.cit., pp. 124 ff. The technical characteristics of the *kobbelbruk* are more thoroughly described below.

better care led to increased yields from the domestic animals as well.[18] The traditional habit of systematic underfeeding of the cattle during winter time, to save as many as possible for the abundant summer pastures, was gradually abandoned with the growth of crop rotation and commercial fertilizers.[19] Tilling of meadow land, fodder cultivation and improved animal care, made it possible to extend the productive period of livestock to the winter season as well. This was particularly attractive if markets were avalaible for regular sale, as will be considered below.

From the middle of the 19th century, there thus developed a less one-sided system of corn production in the eastern and central flatland area, together with an extended production season for livestock in the valley and mountain regions too.

Denmark. The changing of traditional technology and agricultural methods occurred earlier in Denmark than in Norway. The soil and climate were definitely more suitable for agricultural production, and Danish lands were marked by large-scale, market-oriented, semi-capitalist farming during the 18th century (cf. chapter 5). The old intensive one-field system (similar to the *reitbruk* of south-western Norway, and called *alsædesbrug* or *envangsbrug* in Denmark) was, on the whole, already abandoned by the end of the Middle Ages. Towards the end of the 18th century, it prevailed only in some minor areas in north-western Jutland, where excellent pastures made sufficiently intensive manuring possible.[20] The ordinary three-field system, with certain geographically restricted variants, was the basic method on the islands and the eastern parts of Jutland from the Middle Ages to the middle of the 18th century. It implied a rotation between two grain fields and one fallow each year, and was closely connected to the *driftstvang,* the compulsive management system, of communal and semi-communal landholding. The three-field system was replaced by a more flexible 'grass-field system' (*græsmarksbrug*) only in those western areas of Jutland where individual farmsteads prevailed.[21] About the middle of the 18th century, the large manorial farms–which were unrestricted by communal landholding–changed from the traditional three-field system to a kind of convertible husbandry, *kobbel-*

[18] Cf. Borgedal, op.cit., vol. II, Oslo 1967.

[19] Cf. Steen 1957, op.cit., pp. 30 ff. It has been argued that the method of underfeeding was most extremely employed in those areas where the production of *meat* was the dominant aim of livestock husbandry–cf. Stein Tveite, *Jord og gjerning*, Oslo 1959, p. 23.

[20] Cf. K. Hansen, ed., *Det danske Landbrugs historie*, vol. I–V, Gads forlag, Copenhagen 1934–45–here vol. I, p. 96 and vol. II, p. 139.

[21] Ibid., vol. I, p. 96. The *græsmarksbrug* implied a kind of longitudinal rotation, where an area was split into about 8 to 15 fields, which were cultivated continuously for 3 to 5 years each, and then left to grass production–in fact fallow– for 4 to 6 years.

brug, which implied a reduction of fallowing. The basic model of this system comprised eleven enclosed fields, of which five were cultivated with cereals, five with grasses, and one lay fallow–in a more or less regular rotation. The cultivation of fodder crops, and the grazing on the clover fields, raised the level of animal nutrition compared to the mixed grazing of flocks and herds on the wastes and commons of the peasant villages, and made possible a surplus of meat and dairy produce for the market. The large-scale manors (the home farms of the big land-owners) were thus directly integrated into the money economy much earlier than the peasant farms.[22] The peasant lands gradually changed from the rigid three-field system to convertible agriculture parallel to the enclosure movement around the year 1800. This was in fact a diffusion of the traditional *græsmarksbrug* of western Jutland, but–because the soil generally was better in other areas–with shorter periods for grass production between each rotation of creals.[23]

More intensive agricultural techniques developed during the first half of the 19th century. They were stimulated by the substantial population growth (cf. chapter 5), together with a notable improvement in market conditions–especially after the recovery from the depression of the 1820s. Most districts in Denmark were located relatively nearby urban centres and ports of shipment. Technically, the new farming was made possible by enclosure, better-designed implements, and soil improvement.[24]

The development of convertible husbandry, with gradually more flexible rotation of crops, enabled the land to carry more stock, which in turn enriched the soil through heavier manuring. The cultivation of fodder crops like clover also in itself improved the quality of the soil.[25] But the *kobbelbrug,* especially in its earlier phase, still had some definite weaknesses. The relatively one-sided cultivation of cereals for several subsequent years led to an impoverishment of the soil, and fallowing continued to be necessary. Simultaneously, the wastes were increasingly taken in as tillable land–especially from the 1830s. One-sided corn production on the arable land, which by implication followed the one-field or three-field systems, was thus no longer possible, since fodder crops had to be cultivated as a compensation for the former grazing on the *overdrev,* to outer village wastes and scrubs. The gradual adaptation between cereal and fodder crops became a practical

[22] Cf. K. Hansen, op.cit., vol. II, p. 141; and Erling Olsen, *Danmarks økonomiske historie siden 1750,* Copenhagen 1961, pp. 21–23.

[23] K. Hansen, op.cit., vol. II, p. 140.

[24] Cf. ibid., vol. V, p. 203 and p. 207.

[25] Ibid., vol. II, pp. 141 f. In Britain this development seems to have taken place some decades–perhaps about half a century–earlier–cf. J. D. Chambers and G. E. Mingay, *The Agricultural Revolution 1750–1880,* Batsford, London 1966, pp. 54 ff.

argument for a more systematic crop rotation during the second quarter of the 19th century, while it had tentatively been introduced on the market-oriented, large-scale manorial estates some decades earlier.[26] The immediate response in peasant agriculture was a gradual improvement of convertible husbandry, the *kobbelbrug*. The advantages of crop rotation were partially employed through heavier stocking and manuring, together with the cultivation of clover, roots, peas, and potatoes.[27]

Agricultural output expanded more or less continuously during the boom period from 1830 to 1880, and increased yields especially comprised wheat, barley and rye, with a somewhat slower growth of the traditional production of oats.[28] It has been estimated that annual cereal yields were, on the average, about double as high during the 1870s as they had been during the 1830s,[29] in addition to the substantial increase in roots and potato crops well after the middle of the century.[30] The system of real crop rotation without fallowing, however, did not come to dominate peasant agriculture until the breakthrough for commercial livestock husbandry in the last quarter of the 19th century.[31]

The gradual transformation of agricultural practice had a close correlate in the development of farm tools and implements.[32] From late antiquity towards the middle of the 19th century, tilling was usually done by a heavy wheeled plough, pulled by a team of six to eight pairs of horses or oxen.[33] The wheel plough was gradually replaced by a light swing plough pulled by a team of two horses. The innovative forerunners were the large-scale manors–after the abolition of compulsive labour duties from the peasantry. The new swing ploughs, with a wrought-iron mould board, have been estimated as comprising about a tenth of the total number of Danish ploughs about 1820.[34] Iron increasingly replaced wood as the main material for agricultural tools after the depression of the 1830s, and there was a rapid change from self-sufficient production of farm implements to specialized rural

26 Cf. K. Hansen, op.cit., vol. I, pp. 201 f.

27 Ibid., vol. V, p. 208 and p. 287. In academic circles, the system was already being debated at the beginning of the century–cf. ibid., vol. II, pp. 142 ff.

28 Ibid., vol. V, pp. 290 f.

29 Cf. Skrubbeltrang 1938, op.cit., p. 104.

30 Cf. K. Hansen, op.cit., vol. II, p. 145.

31 Ibid., vol. V, p. 299.

32 This is a rough judgement from contemporary estimates, since few accurate figures are available. We have no Danish study corresponding to that of Valen-Sendstad (1964, op.cit.) for Norway.

33 Cf. K. Hansen, op.cit., vol. IV, pp. 429 ff.

34 Ibid., vol. IV, p. 451.

handicraft, and further to standardized manufacture originating in the urban centres.[35]

The wheel plough with four to six pairs of horses was still common in peasant farming about 1840, but it was from now on gradually replaced by lighter wheel ploughs (pulled by a team of two horses), and by English swing ploughs–especially in those regions where the soil was easy to manage. The swing plough was by then predominant on the large-scale farms, where the labour duties had now been abandoned.[36] New iron harrows were introduced into peasant agriculture by the middle of the century.[37] Reapers seem to have replaced the scythe from the mid-1860s, while mowers became usual in the last quarter of the century.[38] The same applies to grain drills, which, however, had been taken into use on the large-scale manorial farms earlier in the century, after the abolition of villeinage.[39] Horse rakes, seeders, and chaffcutters were applied on larger farms from about the middle of the century.[40] The dissemination of threshers had also been counteracted by the Danish system of villeinage when it started little by little in the beginning of the 19th century, and threshing machines became usual on larger farms from the middle of the century.[41] Steam threshers then penetrated Danish agriculture from the last quarter of the century.[42]

We might conclude that the process of technical innovation in Danish agriculture seems to have been more prolonged than in Norway, particularly in this early phase. It started at least half a century earlier, and the original background conditions were also technically less primitive. The development of neo-technical farming in Denmark was marked by ecologically more homogeneous conditions, and by the early market-orientation of the dominant large-scale agriculture, although mechanization here was definitely hampered until the abolition of villeinage. The process slowly penetrated peasant farming parallel to the development of peasant proprietorship in the 19th century. The mercantilization of produce stimulated the peasant urge to freehold, and gradually dissolved the system of leasehold estates (cf. chapter 6).

Sweden. Around the year 1800, Swedish agriculture varied from the technically backward conditions of the Norwegian periphery to the

35 Ibid., vol. V, p. 321.

36 Ibid., vol. IV, pp. 452 ff. This information is based on the county reports of the Agricultural Society–*Landhusholdningsselskabets Amtsbeskrivelser*–which were issued regularly from the middle of the 1820s to the beginning of the 1840s.

37 Ibid., vol. IV, pp. 466 ff.

38 Ibid., vol. IV, pp. 505 ff.

39 Ibid., vol. IV, pp. 483 ff.

40 Ibid., vol. V, p. 322.

41 Skrubbeltrang–1938, op.cit., p. 101–argues that the deficit labour force due to the war of 1848 was a major causal factor.

42 K. Hansen, op.cit., vol. IV, pp. 526 ff. and pp. 531 ff.

relatively advanced methods of manorial farming on the Danish model. The old one-field system, in Sweden called *ensädesbruk,* was still employed in mountainous districts and peripheral areas, while the two- or three-field system dominated the more densely populated regions.[43] The *ensädesbruk* was based on a strict balance between livestock husbandry and arable farming, but this balance was difficult to maintain outside the wooded districts with abundant pasture land, and the soil often became successively exhausted.[44] Particularly, this affected some backward smallholding areas with continuous tilling of the same fields and plots.[45]

The main agricultural areas, however, employed either a regular two-field or three-field system, with one field in a fallow rotation.[46] Even the Norwegian system, with four or more fields in the longitudinal circulation, was employed on certain farms or for some parts of the arable land.[47] The common denominator for all these methods was a permanent separation between fields and meadows. The fields were reserved for cereals, and there was no production of fodder crops. Only some parts of northern Sweden and Dalecarlia knew an irregular kind of circulation, with ploughing and cultivating of former wastes and meadow land.[48]

Peasant agriculture in most of Sweden was also based on the same technical implements as centuries before. The wooded ploughs and harrows were certainly replaced by iron tools on many manorial estates, but until the middle of the 19th century, the wrought-iron plough was unknown in many districts, except mainly Dalecarlia and western Sweden.[49] Like some Norwegian districts, certain regions employed the *ard* –a simple plough without mouldboard. In the southern county of Scania, however, tilling was done–as was the the case on the Danish isles–by the heavy wheel plough with a team of six to eight pairs of oxen or horses.

During the first decades of the 19th century, there was a slow and gradual improvement of agricultural techniques and methods of culti-

[43] Cf. Gustaf Utterström, *Jordbrukets arbetare,* 2 vols., Tidens förlag, Stockholm 1957, vol. I, pp. 487 ff.

[44] Cf. Eli Heckscher 1949, op.cit., Part II, vol. 1, pp. 212–217.

[45] Cf. H. Juhlin Dannfelt, *Kungl. Landtbruksakademien 1813–1912 samt Svenska Landthushållningen under nittonde århundradet,* Stockholm 1913, part I, p. 78.

[46] See E. Ingers, op.cit., vol. II, pp. 112 ff. for details about the regional distribution of major types.

[47] Cf. Juhlin Dannfelt 1913, op.cit., pp. 77 f.

[48] Ibid., p. 79. Cf. the traditional *græsmarksbrug* of western Jutland which was described above.

[49] Cf. Eskerød in Ingers, op.cit., vol. III, p. 15; and Juhlin Dannfelt 1913, op.cit., I, p. 67.

vation.[50] New soil began to be increasingly cultivated, and the reclamation of land expanded particularly from the 1820s, together with the partition of farms–when farmland was still sufficient relative to the population growth. There was also a gradual change from the old field systems to convertible farming and regular rotation of crops. Fodder crops like clover and timothy had been introduced on the large-scale manors in the very beginning of the century.[51] In peasant agriculture, crop rotation was initiated by the lowland farmers of Scania from about the 1820s, and clover was put into circulation from the 1830s–potatoes even earlier. But these methods were not followed by the peasantry of central Oestergoetland until the 1860s, and real crop rotation–or even convertible husbandry–did not penetrate certain areas until still later in the century.[52] The more intensive agriculture from the middle of the 19th century, with consolidation of formerly intermixed fields, crop rotation, and better draining–accompanied by better management of livestock–required improved implements The introduction of iron ploughs and threshers accelerated with the boom period of the 1850s, even if large-scale agriculture had experienced this transformation some decades earlier.[53] Foreign harrows, grain drills and horse rakes were imported towards the middle of the century, and manufacture of agricultural implements, producing for the domestic market, spread shortly after.[54] But the real breakthrough for labour-saving machinery in peasant agriculture occurred during the 1870s and 1880s–with the dissemination of steam threshers, mowers, reapers, chaffcutters, and dairy separators.[55] Artificial fertilizers were increasingly taken into use from the 1860s[56]

It has been thoroughly documented by later research, that the agricultural revolution in Sweden was contingent upon the market situtation.[57] The tempo and scope of modernization of implements then

[50] Utterström argues that the innovations were often atypical, and exaggerated in historical accounts, contrary to Valen-Sendstad's general view about the situation in Norway–cf. Utterström 1957, op.cit., I, p. 444.

[51] Cf. Heckscher 1949, op.cit., II. 1., p. 217.

[52] Cf. Carlsson in Ingers, op.cit., vol. III, pp. 139 ff.; and Juhlin Dannfelt 1913, op.cit., part II, pp. 98 f. Roughly, the change followed the enclosure movement, but occurred *after* the fulfilment of reallotment, and occurred last on peripheral smallholdings.

[53] Utterström 1957, op.cit., I, pp. 658 ff. Utterström on the whole connects agricultural innovations to improved business cycles.

[54] Cf. Juhlin Dannfelt 1913, op.cit., I, pp. 303 ff.

[55] Ibid., pp. 83 ff.

[56] Ibid., pp. 93 ff.

[57] See Jan Kuuse, *Från redskap till maskiner. Mekaniseringsspridning och kommersialisering inom svensk jordbruk 1860–1910*, Göteborg 1970; and Kuuse, 'Mechanisation, Commercialisation and the Protectionist Movement in Swedish Agriculture 1860–1910', *SEHR*, vol. XIX, No. 1, 1971, pp. 23–44. Cf. also the discussion of Norway above.

increased with the size of the farm, and the typically agricultural areas led the way. On the whole, the new farming spread from the southern regions northwards, and from the higher to the lower strata of the agricultural society–from large-scale and market-favoured landlordism to peripheral peasant farming.[58] It has also been argued that the technological gap between large-scale manors and peasant farms in fact *increased* until the 1860s, while the reform period from the 1870s transformed peasant agriculture particularly.[59]

Agricultural credit

The new farming required financial means that surpassed the current accumulation within peasant agriculture itself. The demand for agricultural capital increased with the expansion of land reclamation, field reallotment, and diffusion of technical implements. In principle, this meant no break with the previous economic system. Even earlier financial needs had made a breach in the relatively self-sufficient peasant economy. Freehold purchases, inheritance succession, and enclosures, were principally based on long-term mortgage credit, as we saw in previous chapters. Generally, the monetary requirements of the peasantry increased all over Scandinavia throughout the 18th century, owing to a transition from payments of interests and taxes in kind to payments in money.[60]

The technical innovations in agriculture particularly raised the demand for short-term working capital, but the system of banking and credit remained backward and rudimentary throughout Scandinavia during the first half of the 19th century. The financial needs of the peasantry were principally met by personal credit and usury, although the exact significance of these types of dependence upon arbitrary local power groups is difficult to estimate. The first officially controlled 'credit system' was established with the communal granaries and potato storages, the *bygdemagasin,* from about the middle of the 18th century. These communal storages gave loans in corn or potatoes against interest in kind, and represented primarily an insurance system in case of bad crop failures.

Two types of rural granaries developed in *Sweden* from the second

58 Cf. Kuuse, op.cit.; and Utterström 1957, op.cit., I, p. 445.

59 Cf. Carlsson in Ingers, op.cit., vol. III, pp. 142 f.

60 In Denmark, for instance, the public corn tax had to be rendered in money from 1791, and the regular payment of taxes in kind on the whole ceased to exist from the 1820s. The peasant villeinage on manorial estates was often replaced by monetary duties from the 1790s, while the tithe was gradually paid in money from the first decades of the 19th century, and transformed into a fixed land tax in 1852. Cf. Sigurd Jensen 1950, op.cit., p. 109, p. 112, pp. 121 ff., and p. 149.

quarter of the 18th century–the crown storages (*kronomagasin*) and the parish storages (*sockenmagasin*). The crown storages were partly used to serve military garrisons, partly as stocks of the public taxes which were used for equalization of corn prices, and partly as reserve funds during crop failures.[61] The crown storages declined from about 1830, when the state abolished the remainder barter economy in public finance by abandoning its right to enforce payment of corn duties in kind.[62] The crown granary movement came to a definite end with the complete transition towards monetary public duties from 1869.[63]

The private parish storages were, on the other hand, a kind of primitive non-monetary savings banks, and became the first widespread intermediating association movement among Swedish peasantry.[64] These granaries were initiated by the public authorities, however, and the peasant representatives in the parliamentary assembly were reluctant –apprehending further burdens–when the case was discussed in 1723. The movement did not make progress until the 1750s, when several granary associations were established under public regulations. The storages were based upon corn deposits relative to the fixed rent value of the farms, and gave loans against interest in kind. More than 500 communal storages were established before the year 1800, but the number was fluctuating. It has been argued that their stockholding was insufficient during real crop failures, but they obviously had some importance as institutions for loans of seed and bread grain to the less prosperous strata of the peasantry.[65] The system declined from the middle of the 19th century owing to improved communications and credit institutions, and the country granaries became voluntary communal affairs from the parliamentary session of 1862–63.[66]

While *Danish* granaries were principally organized within the manorial estates, *Norway* had a system comparable to the Swedish one. Several communal granaries were established during the last decades of the 18th century, but they declined rapidly from the 1840s. While there were 216 country storages in 1840, the number was 137 in 1845 and 122 in 1850, and their aggregate stock in hand fell correspondingly.[67]

[61] Cf. Karl Åmark, *Spannmålshandel och spannmålspolitik i Sverige 1719–1830,* Stockholm 1915, pp. 187 ff. and p. 204.

[62] Ibid., p. 292.

[63] Ibid., p. 293. The palace granary of Stockholm was among the very last to be shut down by a royal decree of 1871.

[64] Ibid., pp. 320 ff.; and Ingers, op.cit., vol. II, pp. 250 ff.

[65] Cf. ibid., vol. II, p. 253.

[66] Cf. Åmark 1915, op.cit., p. 341 and p. 346.

[67] Cf. Nicolai Rygg, *Norges Banks historie,* vol I, Kristiania 1918, p. 319; Francis Sejersted, *Norges Bank og høykonjunkturen i 1840-årene,* Universitetsforlaget, Oslo, 1968, p. 24; Steen 1957, op.cit., p. 205. There was no public regulation comparable to the case in Sweden.

The parish granaries in Scandinavia were gradually replaced by private and communal savings banks towards the middle of the 19th century. This development is definitely an expression of the transition from a barter economy towards an economy based on the mediation of money. The rural granaries had given some loans directly in cash in their final phase, and their stock was often sold to constitute the funds of the new savings banks.[68] The development in the three Scandinavian countries seems to have been strikingly parallel. The first private savings bank in Denmark was established in 1810, in Sweden in 1820, and in Norway in 1822.[69] The savings banks originated in the cities, but they were gradually established in the countryside and supplied agricultural credit within the rural communes. Norway had only 22 savings banks in 1840, but the number had increased to 90 in 1850, and further to 174 in 1860 and 262 in 1870.[70] In Sweden there were 60 savings banks in 1850, 86 in 1850, 151 in 1860, and 234 in 1870.[71] This development was in reverse ratio to the decline of communal granaries. Denmark also had nearly 300 savings banks in 1870, while the number had been less than 50 by the middle of the century, and only 76 still in 1864. But it should be noted that local agricultural capital in Denmark was supplied by special 'credit associations' as well. These institutions financed a substantial part of the conversion of tenant farms to freeholds from the 1850s.[72] The earlier Swedish 'mortgage associations' were no substitutes for the communal savings banks, since they reserved the right of mortgage to estates of a minimum assessed value. While the gap between official assessment and real market value steadily increased towards the 1860s, a substantial amount of farm land was excluded from the credit supplies of the mortgage associa-

[68] Cf. Rygg I, op.cit., p. 319; and Steen, *Lokalt selvstyre i Norges bygder,* Cappelen, Oslo 1968, pp. 81 ff. Steen describes how the first savings banks in Norway often also originated as communal institutions in the rural areas.

[69] On agricultural financing and the savings bank movement in Denmark, see K. Hansen, op.cit., vol. V, pp. 327–335. On Sweden, see Emil Sommarin, *Vårt Sparebankväsen 1834–1892,* Lund 1942, and Ingemar Nygren, *Svensk Sparbanksutlåning. 1820–1913.* Gothenburg 1967. On Norway, cf. Steen 1957, op.cit., pp. 214 ff., Rygg I, op.cit., p. 42 and passim. There are also a lot af valuable monographs about individual banks, which are directly relevant to the theme of agricultural financing–see, inter alia, Anne Hals, *Sparebanken og det store hamskiftet. En analyse av Gran Sparebanks utlånspolitikk 1857–1887,* stencilled thesis in history, University of Oslo, 1971. See also the first chapters, by Bjørn R. Rønning and Anne Hals, in *Studier i sparing og sparebankvesen i Norge 1822–1972,* Gyldendal, Oslo 1972.

[70] Cf. Sejersted 1968, op.cit., p. 24; Rygg I, op.cit., p. 42.

[71] *Historisk statistik för Sverige,* Statistisk Centralbyrån, Stockholm 1960, table 83, p. 102.

[72] Cf. K. Hansen, op.cit., vol. V, p. 329 and pp. 332 f. The credit associations also gained importance in Denmark when equal inheritance rights for women were established by law in 1857. A corresponding inheritance law was passed in Sweden in 1845.

tions.[73] The mortgage associations were, however, established in most of the counties of southern and central Sweden between the 1830s and the middle of the century, and were definitely contributing to the remoulding of intermediate and large-scale agriculture.

The initial establishment of savings banks in Scandinavia was not intended to supply short-term credit for agricultural investment. Their purpose was originally paternalistic benevolence: to stimulate saving among the poor.[74] During the first decades loan-giving was based on personal acquaintance and given against a promissory note, while the objectives of professional capital supply became more pronounced during the second half of the century. In Norway, however, the savings banks seldom gave mortgage credit before the last quarter of the century, but short-term working capital was equally important during the agricultural innovation process, and other institutions were established to take care of the long-term financial needs.[75] The new savings banks on the whole became crucial in the short-term marginal financing of the new farming, and they helped to get peasant agriculture out of the hands of private creditors.

The communal savings banks were not, as mentioned, the only source of agricultural credit supply. In Denmark the 'royal credit bank' had supplied agriculture with mortgage capital from 1786 to 1816, and other public institutions did the same throughout the 19th century. The first private business bank was established in Copenhagen in 1857, although its significance for agricultural investment was probably negligible compared to the savings banks and the communal mortgage associations.[76]

Sweden got a system of independent business banks–in addition to the old national bank–from the beginning of the 1830s, and new credit enterprises on a joint-stock basis were established from the 1860s, but these were also more crucial to trade and industry outside agriculture.[77]

Norway also got private business banks from the middle of the century, but most important for agricultural financing was the royal *hypothekbank,* which was established in 1852.[78] This public mortgage insti-

[73] Cf. Nygren 1967, op.cit., p. 28–who argues that this was conducive to the accelerating establishment of communal savings banks from the 1860s.

[74] Cf. Sejersted 1968, op.cit., p. 23; Anne Hals 1971, op.cit.; K. Hansen, op.cit., vol. V; Sommarin 1942, op.cit., p. 106 and passim; Nygren 1967, op.cit., pp. 12 ff., pp. 26 ff. and passim.

[75] The relatively rich peasant farmers, with larger farms, used more advanced agricultural techniques, and had better financial capabilities of investment in a bank. It is documented in a local case-study that these also in fact initiated the new savings bank in the 1850s, cf. Hals 1971, op.cit., pp. 9 ff.

[76] Cf. E. Cohn, *Privatbanken i København gennem hundrede Aar,* Kbh. 1957; and Erling Olsen 1961, op.cit., pp. 281 ff.

[77] Cf. Heckscher 1957, op.cit., pp. 290 f.

tution was founded on the explicit background of the European depression from the revolutionary year 1848, which meant a grave export crisis for the Norwegian lumber trade, and a subsequent credit privation among the peasantry of southern and eastern regions. It has been documented that applications for loans in the *hypothekbank* principally came from those peasant farmers who were relatively most integrated in the money and market economy: peasants on medium-size or larger farms from the eastern and southern counties, and nearly none along the coast from the south-west and northwards.[79] The same application pattern agrees with the effects of the world crisis of 1857, when the commercial interests claimed payment of their accounts outstanding among this market-integrated peasantry, and thus increased the demand for substitute capital from the bank.[80] The same geographical pattern even seems to apply to the extensive transition towards intensive livestock husbandry which followed the depression of the corn market in the 1860s, as the relatively subsistence-oriented peasantry of western and mountainous districts still ignored the bank, or were openly hostile towards it.[81]

We cannot conclude, however, that capital formation as such was absent among a greater part of the Norwegian peasantry—or particularly so in peripheral districts. We know that some improvement in agricultural techniques took place in most parts of the country even during the first half of the 19th century, but intensive local studies are needed to survey possible financial sources alternative to the formal credit institutions. In some Swedish counties there has been documented the existence of a non-institutional agrarian credit market which was financing land reclamation, soil improvement, and technical innovation during the decades up to about 1860.[82] A network of reciprocal financial obligations seems to have made the smallholdings and intermediate holdings agriculturally self-financing, while the more technically advanced largeholdings principally based their financial supplies upon non-agrarian capital and formal credit institutions.[83]

[78] See Alf Kaartvedt and Leif Chr. Hartsang, *Kongeriket Norges Hypotekbank 1852–1952,* Oslo 1952. On the establishment of private banks and insurance companies, see also W. Keilhau, *NFLH,* op.cit., vol. IX, pp. 139 ff. A stimulating analysis of the establishment of one of the first private joint-stock banks in Norway is presented by Jens Arup Seip, 'Saint-Simon, Credit mobilier og Den norske Creditbank', *Historisk tidsskrift,* vol. 36 (1953), pp. 513–538—also published in Seip, *Tanke og handling i norsk historie,* Gyldendal, Oslo 1968.

[79] Cf. Kaartvedt and Hartsang 1952, op.cit., pp. 107 ff.

[80] Ibid., pp. 114 f.

[81] Ibid., p. 136 and pp. 144–151. This attitude is, as we shall see, crucial to our analysis in chapter 9 below.

[82] Sture Martinius, *Agrar kapitalbildning och finansiering 1833–1892,* Gothenburg 1970.

[83] Ibid., pp. 137 f., pp. 147 ff. and passim.

This pattern may indicate that the earlier penetration of large-scale Swedish agriculture by the institutionalized financial market was dependent upon an economic surplus derived from manufacture and commercial activities.[84] In Denmark, large-scale agriculture itself seems to have been the main source of capital formation at the national level, while fishing, the lumber trade, and shipping counted heavily in early Norwegian-capital formation.[85]

None of the Scandivavian countries established an institutionalized credit system until about the middle of the 19th century. The relatively large-scale agriculture in geographically central areas was first able to innovate with financial loans from modern banks, while the peripheral and small-scale farming principally remained outside this financial penetration until well into the second half of the century. This applies particularly to the peasantry of substantial parts of Norway and northern Sweden. The growth of modern credit institutions in agriculture implied a centripetal force which broke the power of parochial usury.

The stable mercantilization of agriculture does also require an ordered credit system, since the risk of market dependence—as well as the investment needed—demands a predictable supply of capital. The first half of the 19th century saw a gradual transition from communal granaries, based on deposits and loans in kind, to a credit system adapted to the growing market and money economy, and from personal dependence and usury towards standardized public criteria of loan-giving and interest. This was an important aspect of the penetration of land by the market system—through a gradual integration of the peasantry within a nation-wide financial framework.[86] The corresponding commercial penetration, which to some extent constitutes the background of this financial system, will be discussed in the following section.

The intrusion of the market

There seems to have been a slow but unambiguous increase in the peasants' need of cash during the 18th and early 19th centuries all over the Scandinavian area. Public obligations gradually changed from duties in kind and in labour to taxes that were to be rendered in money.

[84] Cf. Heckscher 1949, op.cit., II. 1., and Heckscher 1957, op.cit.

[85] This difference is also suggested by Lennart Jörberg, *The Industrial Revolution in Scandinavia 1850–1914,* Fontana Books 1970, p. 27. C. Svend Aage Hansen, *Early Industrialization in Denmark,* Copenhagen 1970.

[86] It should be noted that the communal savings banks were not completely localistic institutions. They could borrow from other banks in case of capital shortage, and the rates of interest were regulated by law.

The population growth, which accelerated during the early 19th century, also broke the relatively stable framework of traditional land husbandry. The new farming, and the credit supply upon which it was based, went hand in hand with a gradual extension of the market for agricultural produce. The increasing urbanization, and the improvement of communication and transport, indicates the time sequence of this process.

The domestic market for agricultural produce was immediately stimulated by the growth of a non-agrarian urban population, and improved means of transport furthered the commercialization of farming in wider circles outside the close neighbourhood of towns.

Urbanization. Even during the 18th century, urbanization was relatively advanced in Denmark as compared to Norway and Sweden, as we saw in chapter 5. The proportion of urban dwellers relative to the total population also remained quite stable during the first half of the 19th century. While the urban population in Denmark accounted for slightly above 20 %, it only amounted to about 10 % of the total population in Norway and Sweden. Then the process of urbanization accelerated in all the Scandinavian countries during the second half of the century–slowly in the first two decades and then faster from about 1870. The general picture is indicated by the following table:

Urban dwellers in percentage of total population[87]

Year	Denmark	Norway	Sweden
1800/01	21.1	8.8	10.0
1850	–	–	10.1
1855	22.6	13.3	–
1870	24.9	–	13.0
1875	–	18.1	–
1900/01	39.2	28.0	22.0

Source: The official censuses. Here from A. S. Svalestuen 1971, p. 18.

[87] The accuracy of these figures should not of course be taken too literally, since we have no guarantee against errors or non-uniform definitions of 'urban area' from census to census. It is also very likely that the relatively high proportion of urban dwellers in Norway relative to Sweden is due to different structures of urbanization, with a great number of rather small cities in Norway–particularly along the southern coast. This might be cleared up with comparative figures for the amount of urban dwellers in agglomerations above a certain size; but not even the small towns were populated by peasants, however. So the major trend is sufficient for our purpose, even if the comparative figures might be somewhat misleading. Cf. also *Historical Statistics 1968*, Oslo, table 13, p. 33.

This development indicates that the substantial population growth from 1820 to 1870 was principally absorbed in the countryside, even if the towns expanded markedly in absolute number of inhabitants. The clear *relative* extension of the non-agrarian urban population occurred from about 1870, parallel to the beginning of industrialization in the North. Even earlier, however, there occurred a gradual occupational differentiation of the rural periphery–as indicated by a population growth in the more densely populated areas within rural municipalities.[88]

Communications. It would be misleading to make an unreserved comparison of the development in the transport and communication systems, since distances and topographical conditions varied greatly between the three countries. In flatland Denmark, most of the land had easy access to nearby towns. Road construction was fairly easy, and represented no crucial barrier against commodity transport and trade. This was not so in substantial parts of Norway and Sweden, where many isolated rural communites–and not only the most remote areas–were more or less blocked off from the national communication systems. A certain improvement of central main roads took place in Norway in the decades after the separation from Denmark, and some of the bigger mountain valleys got permanent road connections during the 1830s and 1840s.[89] From 1840, the development of the *length of public roads* runs as follows:

1840	15 290 kms	1870	20 058 kms
1850	16 091 kms	1880	22 194 kms
1860	17 933 kms	1890	25 484 kms

Source: *Historical Statistics* 1968, Oslo, table 205, p. 415.

In addition to this development, there was a notable improvement of the standard of existing roads, even if substantial parts of the peasantry remained well outside the integrating network. In western parts of the country, with deep fjords and steep valleys, water transport was dominant throughout our period.

Sweden experienced some improvement in overland communications from the 1830s, when new areas were opened up for an expansion in trade.[90] Then from the 1850s there was a notable increase in public road construction, and the improved overland communications also

[88] See *Historical Statistics 1968,* Oslo, diagram p. 28.

[89] *Hallingdal* got its main road in the 1830s, and *Setesdal* in the 1840s–A. Bergsgård, *Norsk historie 1814–1880,* 2nd ed., Det Norske Samlaget, Oslo 1964, p. 166. Cf. O. Kristiansen, *Samferdsel i Norge 1814–1830*, Oslo 1926.

[90] Cf. Åmark 1915, op.cit., p. 59 and passim.; Utterström 1957, op.cit., pp. 393 ff. The earlier public engagement in communications had concentrated on ports and canals.

meant–as in Norway–that the peasant duties of transport and traveller conveyance were reduced, so it became easier to concentrate on farming.[91]

The Danish road system was slowly improved after the draft of a national plan in 1793, but real progress had to wait until the 1830s, and the planned network was effectively implemented during the 1860s.[92]

The transport facilities in both Norway and Sweden made a great leap forward with the development of railroads from the middle of the 1850s.[93]

Total length of railway lines in Norway and Sweden (kms)

Year	Norway	Sweden
1855/56	68	66
1860	68	527
1870	359	1727
1880	1115	5876

Sources: *Historical Statistics 1968,* Oslo, table 200, p. 410 and *Historisk statistik för Sverige,* Statistiska Centralbyrån, Sth. 1960, table 44, p. 76.

The public railway construction reached considerable proportions in both countries during the 1870s, even if the Swedish acceleration started earlier and showed a boom that exceeded the Norwegian one. There was a demand from the growing export industry for improved transport from inland to the ports. In Sweden this included grain crops, while timber, wood and pulp were probably the main commodities on the early Norwegian trains.[94] But the new railroads also facilitated the transition towards commercial dairy farming from the 1860s and 1870s. The first Danish railway was built in the 1840s, to serve the trade with Hamburg, while the length of lines particularly expanded

91 Cf. Utterström, 1957, I, pp. 403 ff. and p. 433.

92 Cf. Erling Olsen 1961, op.cit., pp. 248 f.

93 The Swedish public policy and debate on early railroads is outlined by Heckscher 1957, op.cit., pp. 280 ff. The particular Norwegian mixture of private initiative and public support is analysed by Jens Arup Seip, 'Det norske system i den økonomiske liberalismes klassiske tid (1850–1870)' in *Tanke og handling i norsk historie,* op.cit.

94 Cf. Lennart Jörberg, 'Structural Change and Economic Growth: Sweden in the 19th Century', *Economy and History,* vol. VIII, 1965, p. 17. Cf. also Jörberg 1970, op.cit. A critique of the account of Norway in the last work is put forward in a review article by Francis Sejersted, 'Comment. The Industrial Revolution in Norway', *SEHR,* vol. XX, no. 2, 1972, pp. 153–164–with a reply by Jörberg and a rejoinder by Sejersted.

through public initiative during the 1860s. Danish railways formed 1100 kms in 1874 and about 3000 kms in 1900.[95]

Agricultural trade. The dominant types of trade with agricultural goods within each of the countries were transformed during the 19th century. There was a general development from barter to monetary trade, from part-time pedlars and seasonal fairs towards specialized trademanship and stable relations with an impersonal network market.

The barter economy of the Norwegian peasantry during the 18th century expressed the need for the necessities of life like salt, iron and metals, and the need for bread grain after crop failures or in districts where livelihood was based upon animal husbandry or fishing. There was also a modest demand for 'luxury articles' like sugar, tobacco or coffee. Thus not even the most peripheral peasant communities were completely self-sufficient. But barter and trade were basically subsistence-oriented, and did not aim for profits or the reinvestment of commercial incomes. Buying and selling were often mediated by small-scale middlemen–cattle dealers and travelling pedlars–with seasonal trademanship as an occupation subsidiary to peasant farming.[96] Towards the end of the 18th century, the exchange of goods was increasingly mediated by money, and the trend towards specialization became more pronounced. But peasant markets and trade remained, principally, barter transactions–often based on mutual credit-giving through semi-professional middlemen–during the first half of the 19th century as well. The main routes of peasant trade were conditioned by the regional variations between districts based on forestry, fishing, arable farming or animal husbandry. The trade aimed at improvement of the subsistence level, and the orientation towards the agricultural market was *ad hoc* and variable: during the bad years preceding the 1830s, both animal husbandry and the production of cereals aimed at a greater degree of self-sufficiency, while the volume of trade seems to have been temporarily reduced.[97] After the middle of the 19th century, there was a definite change in the relationships of trade. Country tradesmen settled permanently in many counties–especially from the 1860s and 1870s–and sales conditions with less seasonal variations developed for animal produce like milk and butter. Commercial agriculture became a source of reinvestment for agricultural improvement, as a response to population growth and to public obligations, which

[95] Cf. *Danmarks Historie,* Politikens forlag, Copenhagen 1964, vol. 11, '1830–1870' by Roar Skovmand, pp. 372–381; and Erling Olsen 1961, op.cit., p. 249.

[96] Cf. Borgedal, vol. III, op.cit., pp. 105 ff.; Steen 'Bisysler for bøndene' 1933, op.cit., pp. 270 ff.; Steen 1957, op.cit., pp. 43–49; Tveite 1959, op.cit., pp. 31 ff., p. 72 and passim.

[97] Cf. Tveite 1959, op.cit., p. 93.

were to be increasingly rendered in money. The expansion of a non-agrarian urban market supplied the precondition for this response.

The internal peasant trade in *Sweden* was also a barter activity well into the 19th century.[98] There was a rather extensive grain trade between excess and deficit areas, as well as deliveries to the capital and to mining centres in northern and central Sweden.[99] It was particularly the big land-owners who produced a grain surplus for regular market sale, while most of the peasant surplus–outside the most fertile and prosperous districts–went to duties, rents and taxes.[100] As in Norway, the exchange of goods often took place at seasonal fairs, or was mediated by travelling pedlars. In the central wooded districts and northwards, animal produce like butter, cheese and skins were exchanged against other necessities of life.[101] It has been documented that the large-scale manorial farming was a market-integrated enterprise, and a stable customer of non-agrarian goods for farm management as well as for consumption, as early as about 1830.[102] Peasant family farming was, on the other hand, still a subsistence-oriented household economy with limited production for the market, and with exchange of goods supplying immediate personal needs.[103] Peasant farming does not seem to have become more thoroughly market-oriented, with determined reinvestment of commercial profits, until the second half of the century.[104] This was also expressed in a marked expansion of grain trade within the country.[105]

The big land-owners were the principal agricultural operators on the vertical market in *Denmark* around the year 1800. The exchange of goods mediated by money was rather modest between peasantry and tradesmen, and peasant demand was principally limited to the necessities of life.[106] But in Denmark the move towards peasant proprietorship may in itself be seen as a transition towards a money and market economy, and it also went hand in hand with an extension of market trade towards the middle of the century. Danish freehold was as such a leap towards direct market integration of the peasantry–unrestricted by the paternalistic mediation of the landlords–and implied a transition from personal dependence to small-scale enterpreneurship facing the

[98] Cf. Åmark 1915, op.cit., pp. 38 ff.

[99] Ibid., p. 35.

[100] Ibid., pp. 35 f.

[101] Cf. Utterström 1957, op.cit., I, p. 508.

[102] Sture Martinius, *Jordbruk och ekonomisk tillväxt i Sverige 1830–1870,* Gothenburg 1970, p. 35, p. 37 and passim.

[103] Martinius, op.cit., and idem, *Agrar kapitalbildning och finansiering*, op.cit., pp. 41 f.

[104] See also Heckscher 1949, op.cit., pp. 231 ff.

[105] Cf. Utterström 1957, op.cit., I, pp. 221 f.

[106] Cf. Jensen 1950, op.cit., pp. 151 ff.

forces of the market.[107] Danish agriculture had geographically more favourable sales conditions than Norwegian or Swedish, as it was situated at a shorter distance from towns and ports of trade, and there was a quite steady agricultural market boom–which favoured arable farming–for nearly half a century after the depression of the 1820s. This was later called the great 'period of grain sales', and occurred parallel to the definite breakthrough for peasant freehold, as we saw in chapter 6.[108] The peasantry, which even earlier engaged in occasional sales of grain, victuals, turf and wood to the urban centres, now became better customers for commercial goods according to the improvements in farming.[109] This development of more intensive methods of market-oriented farming thus also implied an 'agriculturalization' of the countryside, with a notable decline of rural home crafts.[110]

Towards liberalism in trade and crafts. The development of commercial agriculture in Scandinavia was facilitated by the removal of institutional barriers like bourgeois privileges of trade and public duties on deliveries to the towns. The mercantilistic policy of trade was gradually abandoned–from the first signs of liberalism in the last decade of the 18th century to the final abolishment of guilds and trade restrictions after the middle of the 19th century. The time sequence of the process was fairly parallel all over Scandinavia, and the economic liberalism was also a uniform peasant demand–as we shall see in the next chapter –which facilitated the direct access to the vertical market and thus stimulated the commercialization of agriculture. The Danish monopoly on grain export to southern Norway was abolished in 1788, together with the privileges for landlords and urban tradesmen on trade in cattle within Denmark.[111] In Norway the general trade privileges of the towns were first confirmed in a law of 1818. Commercial legislation in 1842 then liberated the exercise of trade within the cities, while retaining rural trade in the hands of authorized commerce within a certain circumference around the towns. Commerce was made totally free within this circumference in 1857. The remaining restrictions on country trade were abolished in 1866, including the prohibition on import directly from abroad.[112] The sawmill privileges were abolished

[107] Ibid., p. 151 and p. 161.

[108] Cf. K. Hansen, op.cit., vol. V, p. 207, p. 229 and pp. 290 f.

[109] Cf. Skrubbeltrang 1938, op.cit., pp. 156 f.

[110] Ibid., pp. 116 f. Rural home crafts prevailed longest on Jutland–compared to the economically more 'advanced' islands.

[111] Cf. *Danmarks Historie*, 1964, op.cit., vol. 10, 1784–1830, by Jens Vibæk, pp. 105 ff.

[112] In 1882, country tradesmen also got the right of free export, and the last commercial discrimination against the countryside thus disappeared. This had been a demand from the peasant representatives in the *Storting* (Parliament) during the 1870s.

in Norway in 1854 (cf. chapter 5), while the guild system legally disappeared from 1839.[113]

Freedom of trade and crafts in Denmark was introduced by law in 1857, when the guild system was given a transitional period of four years within which to dissolve, and the peasantry acquired a right of trade outside the privileged towns and authorized market places.[114]

In Sweden grain trade was liberated in the last quarter of the 18th century, while free trade in cattle and victuals followed in 1815.[115] The general urban commercial privileges were then abolished in 1846, together with the guild system; and certain remaining minor restrictions were removed in 1862–63.[116]

The transformation of Scandinavian agriculture around the middle of the 19th century thus took place within a new commercial setting based on the principles of economic liberalism. The agrarian interests were, as we shall see later, pressing harder for this policy parallel to the increasing need of production for the market, while the policy in turn facilitated the market integration of the peasantry.

Agricultural export. The agricultural difference between the three countries definitely becomes more visible if we look at the pattern of grain exports. While Swedish and Danish farming produced a notable surplus for export–particularly in oats–from the middle of the 19th century, Norway was not even able to supply her own non-agrarian population.[117] Norway's foreign earnings were based on shipping, the lumber trade, and fishing/whaling.[118] The commercialization of Swedish

113 This was already decided by a majority of the *Storting* in 1827, but the King exerised his right of refusing sanction.

114 The towns got a circumference of less than a mile, outside which trade was free. The mercantilistic policy had originally favoured Copenhagen at the expense of other towns, but all market towns were made formally equal after the new customs law of 1797. Cf. *Danmarks Historie*, op.cit., vol. 10, pp. 105 ff., and Erling Olsen 1961, op.cit., p. 119 and p. 246. On 19th century commercial legislation, cf. Erling Olsen, ibid., pp. 127 ff. and p. 247; and *Danmarks Historie* vol. 11, op.cit., pp. 361 ff.

115 Grain trade was liberated in the more fertile districts by a decree of 1775, and extended to the rest of the country in 1780. The bourgeois representatives opposed this 'reform-mercantilistic' legislation. Cf. Åmark 1915, op.cit., pp. 92 ff. and p. 115; Ingers, op.cit., vol. II, pp. 332 ff. and vol. III (Carlsson), p. 162; Heckscher 1949, op.cit., 235.

116 Cf. Carlsson in Ingers III, op.cit., pp. 162 ff.; Åmark 1915, op.cit., pp. 101 f.

117 Cf. Jørberg 1970, op.cit., pp. 37 f. The Norwegian export of oats during the 1850s and 1860s was extremely modest, and heavily outweighed by the substantial import of bread-grain like wheat, barley and rye throughout the century–cf. *Historical Statistics 1968*, Oslo, table 160, pp. 296 ff., and table 156, pp. 270 ff. Already about the year 1800, 40 % of Norway's grain consumption is said to have been imported–Jørberg 1970, pp. 36 f. But the statistical basis of this estimate is extremely dubious.

118 Shipping accounted for more than 40 % during the 1860s and 1870s, fishing and whaling products for about 20 %, and timber, wood, pulp and paper also for about 20 %–cf. Jörberg 1970, op.cit., table 22, p. 69, and Juul Bjerke, *Langtidslinjer i norsk økonomi 1865–1960,* Statistisk Sentralbyrå, Oslo 1966, p. 64.

and Danish agriculture was, on the contrary, stimulated by a swift expansion of grain exports during the 1850s, which also increased the amount of capital for farm improvements.[119] Sweden's major export items during the decades after 1850 were oats, together with timber and iron, while oats alone were heavily dominant in Danish exports.[120] In both countries the process seems to have started with the agricultural price boom from about 1830. From then on, Sweden–on a small scale at first–produced a grain surplus for export. This development accelerated notably in the 1850s, and reached a maximum in the 1870s.[121] Danish agriculture was favoured by easy access to the ports of trade, and her grain exports were also stimulated when the British embargo against foreign grain was replaced by a more flexible system of customs regulation in 1828.[122] The expansion during the 1850s–like the parallel development in Swedish exports–occurred after the final breakthrough for free trade in Britain in 1846, and increasing prices also invited peasant agriculture to produce for the external market. But even Danish grain exports stagnated during the early 1870s, owing to increased competition from Russian and American grain. The subsequent changeover to animal husbandry and livestock rearing all over Scandinavia will be shortly discussed below, but it should be noted that both Denmark and Sweden–and Denmark in particular–expanded the foreign market for meat and dairy produce, although both countries became net importers of grain.[123]

Mercantilization of land. The commercial mobilization of landed property, as expressed by the development of sales and prices of agricultural holdings, increased rapidly in the decades around the middle of the 19th century.

In Norway, the average price on real property in the rural areas had been fairly stable from the 1830s to the middle of the 1840s. The notable increase–which on the whole started in this decade–experi-

[119] Cf. Utterstrøm 1957, op.cit., I, p. 448 and p. 717; Skrubbeltrang 1938, op.cit., p. 100.

[120] Jörberg 1970, op.cit., p. 39 and passim.

[121] Cf. Gunnar Fridlizius, 'Sweden's Exports 1850–1960', *Economy and History*, vol VI, 1963, chart on p. 13; Jörberg 1965, op.cit., pp. 3 ff. and pp. 14 f. *Oats* quickly became the dominant cereal in large areas of central and southern Sweden during the 1850s–exported mainly to Great Britain. Fridlizius, op.cit., p. 17, estimates also that about 40 % of the entire harvest of oats must have been exported at the beginning of the 1870s. Potato growing also expanded markedly from the 1850s, and the changeover from grain consumption to potato consumption probably freed grain for export. Jörberg 1970, op.cit., p. 39.

[122] The system implied high protective customs when prices on the home market were low, and vice versa. On the effects on Danish agriculture, cf. Erling Olsen 1961, op.cit., pp. 48 f.

[123] Cf. Fridlizius 1963, op.cit., p. 18, and Jörberg 1970, op.cit., p. 28. Norway's exports of condensed milk and butter from the 1870s and 1880s were, by comparison, rather modest–cf. *Historical Statistics 1968*, Oslo, table 161, pp. 298 ff.

enced a temporary setback after the general crisis of 1848.[124] The average price of agricultural holdings–per unit of assessed site value (*skuldmark*)–increased rapidly again in the middle of the 1850s, but then stagnated with the grain depression towards the end of the decade, and showed a slight decrease during the 1860s. Prices then rose again towards the middle of the 1870s, and then remained fairly stable throughout the century.[125] Prices on landed property on the whole doubled from 1822 to 1866.[126]

The prices of agricultural property in Denmark increased from about 1830, from a very low level at the end of the depression of the 1820s. The increase was modest at first, but accelerated in the beginning of the 1840s, and was continually on the up-grade–with varying rates–to the middle of the 1870s, when the market price of grain fell considerably.[127] It should be noted that the prices were, on the average, higher for freeeholds on the islands relative to Jutland, even if prices here increased at least as rapidly during the 1850s and 1860s.[128] The average prices per unit of assessed value (*hartkorn*) were originally substantially higher on larger holdings, but the relative increase in prices on small and intermediate farms accelerated considerably faster from the 1840s to the 1870s, and the price level per unit of assessed value became quite even for all groups of farm size towards the end of the period.[129] The 'grain sales period' implied–gradually–the market-integration of smaller holdings as well.

Even in Sweden the price of agricultural property increased from the 1830s, to accelerate from the middle of the century.[130] The vast increase in the value of landed property was reflected in the official assessment revision of 1862, which then also–as we shall see in the next chapter–qualified a majority of land-owners for the vote.[131] There seems to have been an increasing amount of free farm purchases as well–outside family succession–during the decades around the middle of the 19th century.[132] This trend was probably stimulated by the annulment of the legal remnants of familial landholding after the middle

124 Cf. *Historical Statistics 1968*, table 99, pp. 149 f.

125 Ibid., diagram 24, p. 130.

126 *Statistiske Oversikter 1926*, Oslo, table 16. Cf. also Valen-Sendstad 1964, op.cit., p. 275, and Simen Skappel, 'De høie priser paa jordeiendomme', *Statsøkonomisk Tidsskrift* 1916.

127 Cf. K. Hansen, op.cit., vol. V, pp. 323 ff. There was a stagnation after 1848, and from about 1857. The increasing prices on land were also substantial relative to the general rise in prices–ibid., p. 327. Cf. also A. Jensen, 'De høje Priser paa Landejendomme', *Nationaløkonomisk Tidsskrift*, 1912.

128 K. Hansen, op.cit., vol. V, p. 324.

129 Ibid., table on p. 325.

130 Cf. Martinius, 'Agrar kapitalbildning . . .', 1970, op.cit., pp. 50 ff.

131 Ibid., pp. 156 f.

132 Ibid., table on p. 60 for the three communes of the study.

of the century. The system of 'inheritance land' (*arvejord*)–implying that certain relatives could retract land that was bequeathed or given away–was abolished in 1857, and the *bördsrett*–stipulating that relatives had a right of preemption against the actual market price–was annulled in 1863.[133] The peasant proprietors on the whole opposed this abolishment of the remnants of kinship landholding, while the new laws favoured the landless as well as the bourgeois investors. The mercantilization of land was also reflected in the marked increase of agricultural mortgage indebtedness from the beginning of the 1860s, accompanied by increased prices after foreclosures and rural bankruptcy.[134] The new possibility of exports of grain from southern and western Sweden also stimulated the increasing value of land as such.[135] This general connection is confirmed by the Norwegian case, where those holdings that were situated in market-favoured areas, and thus relatively well integrated in the money economy, also showed the highest prices on the market for land.[136]

Market sale and methods of farming. At several points in the preceding outline, we have suggested interpretations which make the new farming an expression of market adjustment. This connection can now be specified more explicitly on the basis of empirical evidence. Large-scale manorial farming in Denmark and southern Sweden, and bourgeois 'proprietor farming' in south-eastern Norway, were initiating commercial agriculture several decades ahead of the peasantry. Their role as pioneers for the new farming was–particularly in Sweden and Norway–originally financed by a commercial surplus derived from mining, lumber, and trade. The subsistence-oriented peasantry could not afford the risk of adventurism until orderly relations of sale and credit supply made innovation less daring, while the demographic development also made that reorientation more urgent. There is a functional interconnection between market relations and methods of farming. Relatively stable and predictable market sales, with an adequate supply of commercial fertilizer, is a prerequisite for the profitableness of systematic crop rotation. The agricultural adaptation is then conditioned by *economic relationships* with generalized institutions

[133] Cf. Ingers, op.cit., vol. III (Carlsson), p. 212 and Olle Gellerman, *Staten och jordbruket 1867–1918,* Almqvist och Wiksell, Stockholm 1958, p. 9.

[134] Cf. Gellerman 1958, op.cit., pp. 9 ff.

[135] Cf. Martinius, 'Agrar kapitalbildning . . .', 1970, op.cit., p. 158, and Martinius, 'Jordbruk . . .', 1970, op.cit., pp. 32 f.

[136] Cf. Valen-Sendstad 1964, op.cit., p. 275. This also corresponds to the geographical pattern of applications for credit from the *Hypothekbank* after 1859, as described above. In Norway, however, the freeholding peasantry were strong enough to resist the attacks on their old allodial defence institution, the *odelsrett,* around the middle of the century. This definitely restricted the direct commercial mobilization of landed property.

outside the peasant community. Ordered crop rotation, with cyclical tilling of meadow land, demands investment in implements that are beyond the capacity of self-sufficient tool-making. The more simple one-, two-, or three-field systems, with different balance ratios to natural manure supply, are adapted to a greater degree of subsistence production. Economic relationships, with considerable supplies of equipment from outside, are less decisive, while the mode of production is more contingent upon conditions of topography, soil and climate. The dissolution of communal landholding through enclosure and reallotment is also a prerequisite for individual adaptation to the market, since the demand for a greater surplus beyond subsistence production is increasing. Consolidation of individual holdings then reduces the potential for conflict over production as well as circulation of commercial goods. The mercantilization of peasant agriculture generally stimulated productivity in farming, both in absolute terms and in yields per acre, and it was in turn stimulated by it.

The peasant family farming in Sweden and Denmark seems to have experienced a considerable growth of market integration from the 1830s and 1840s, while large-scale agriculture already produced a considerable share for market sale during the first decades of the century.[137] Peasant farming in Norway remained chiefly subsistence-oriented until well into the second half of the century, although the process of increasing market sales definitely started earlier in many areas. The 1860s may nevertheless be said to constitute a threshold in a process which still goes on in some peripheral districts.

Transition towards livestock husbandry. The 1860s were, in a specific sense, crucial not only for the development of Norwegian commerical agriculture, but also for that of Sweden and Denmark. The general transition towards crop rotation and mixed farming accelerated after the falling market price on corn from about 1860. Improved transport systems stimulated imports of cheap foreign grain, and Scandinavian agriculture responded with a changing market adaptation, implying greater emphasis on livestock rearing, since terms of trade were relatively improved for animal products. The falling grain prices were beneficial to those districts that *had* to buy corn, but meant a retrogression for the relatively specialized grain areas. In Norway this setback particularly hit the corn districts of the east and of northern Gudbrandsdal, which also lost their hegemony on deliveries to nearby im-

137 Cf. Martinius, 'Agrar kapitalbildning . . .', 1970, op.cit. Still in the 1860s, however, only a half of large-scale agricultural produce is estimated as being sent to the market in those three communes in the county of Skaraborg which are included in Martinius' study. But in peasant family farming this share was substantially smaller–ibid., pp. 41 f.

Total number of dairies and cheese factories in Norway[138]

Year	1860	1870	1880	1890	1900
	7	39	131	308	780

Source: Knut Dahl Jacobsen, 1964, p. 165.

porters, owing to improved communications.[139] The grain areas compensated by mechanization, fodder crops, and a change-over to milk products. This applied even to the traditional areas of livestock husbandry, since milk products were less exposed to competition from foreign imports than meat and skins. Urbanization and better transport expanded the domestic market demand for milk, butter and cheese. Danish agriculture also improved animal husbandry from the 1860s. The number of horned cattle increased by nearly a third during the 1860s and 1870s, and yields of butter doubled during that period.[140]

The same pattern applies to Sweden, where peasant agriculture in-

[138] The details of the development, county by county, run as follows:

County	1860	1870	1880	1890	1900
Smaalenene	–	5	9	23	24
Akershus	1	4	8	27	50
Hedemarken	1	3	17	26	26
Kristians	–	10	25	40	31
Buskerud	1	5	15	34	43
Jarlsberg & Larvik	–	6	21	44	53
Bratsberg	–	1	4	6	21
Nedenes	–	–	–	2	11
Lister og Mandals	–	–	–	3	18
Stavanger	2	1	4	14	22
Søndre Bergenhus	–	–	1	9	72
Nordre Bergenhus	–	–	–	4	65
Romsdals	–	–	–	4	153
S. Trondhjems	–	1	9	27	53
N. Trondhjems	2	3	18	44	76
Nordlands	–	–	–	1	59
Tromsø	–	–	–	–	3
In all	7	39	131	308	780

Source: KDJ, 1964, p. 165–cf. S. J. Benterud, J. Grude og E. E. Wold, *Meieridriften i Norge indtil aaret 1905*, Kristiania 1908, p. 106.

[139] Cf. Tveite 1959, op.cit., pp. 248 ff.

[140] Cf. Skrubbeltrang 1938, op.cit., p. 105, and K. Hansen, op.cit., vol. V, pp. 298 ff. The manor farms were still, however, leading in the field of dairy management, owing to better storage and treatment of milk and butter.

novated rapidly from the middle of the 1860s, and turned to ordered crop rotation, dairy farming and exports of butter, as more stable market relations expanded.[141]

Dairy farming. The development of commercial livestock husbandry is conspicuously expressed by the growth of dairies and cheese factories during the last decades of the century. The first co-operative cheese factories in rural Norway originated in the middle of the 1850s, and the number increased gradually during the 1860s.[142] (See table at top of p. 181.)

It has been estimated that yields of milk per cow increased by a third from 1865 to 1900.[143] The surplus of produce available for market sale was thus based on improved management of livestock, since the total number of cattle decreased slightly during the same period.[144]

It would be misleading to present corresponding figures for the growth of dairy farming in Sweden and Denmark. Private manorial dairies had existed here for several decades, and after the middle of the century some of them gradually acquired a semi-public character, whereby the peasantry were allowed to separate their milk, churn their butter, and make their cheese before deliveries to the market. In Denmark the first co-operative peasant dairy (*fællesmejeri*) was established on the island of Funen in 1863, and several others followed in Zealand during the latter part of the 1860s.[145] Until 1880, however, there was an official price-discrimination favouring butter produced on the manor farms (*herregårdssmør*) against butter produced by the peasantry (*bøndersmør*), because of the original difference in quality of produce. The number of co-operative dairies now increased rapidly during the 1880s, and numbered in all 680 in the year 1890.[146]

Commercial production of milk in Sweden was mainly restricted to the large-scale manors of the south until the middle of the 19th century. The dairies established during the second half of the century were private manorial ones (*herregårdsmeierier*), and later wholesale commercial institutions (*oppkjøpsmeierier*).[147] Rural dairies on a peasant co-operative basis developed slowly during the 1880s, and accelerated during the last decade of the century. But contrary to the situation in

141 Cf. Ingers, op.cit., vol. III (Carlsson), pp. 158 ff., p. 332 and pp. 350 ff.

142 Cf. Kåre Lunden, 'Framveksten av samvirketiltak mellom norske melkeprodusenter i tiden 1856–1905', stencil, Oslo 1970. The new dairies get a semi-official support from the *Selskapet for Norges Vel* (see below)–cf. Borgedal vol. III, 1968, op.cit., pp. 132 ff.

143 Knut Dahl Jacobsen 1964, op.cit., pp. 139.

144 A certain export of butter and condensed milk took place from the 1880s, while imports of pork and meat on the other hand expanded considerably–ibid., p. 140.

145 According to Skrubbeltrang 1938, op.cit., p. 101.

146 Ibid., p. 111.

147 Cf. Ingers, op.cit., vol. III (Eskerød), pp. 76 ff.

Norway and Denmark, where co-operative institutions dominated the marketing of milk produce, the co-operative ones were in a minority in Sweden: they constituted only 75 out of a total number of 1562 dairies in 1890, and 430 out of 1688 in 1900.[148]

Rural co-operatives. The real breakthrough for commercial agriculture occurred during the last decades of the 19th century. The details of this process, after the initial phase before the 1870s, is outside the scope of the present study. We have intended to outline the original character of the transformation, while the fulfilment, in width and quantitative terms, belongs to a later era. But some national characteristics of peasant organization for the marketing of produce should be noted, since they also retrospectively highlight the non-uniform character of Scandinavian agrarianism. The most notable trait in Danish peasant farming during the last two decades of the 19th century, was the co-operative movement. The co-operatives were established not only for processing and circulation of dairy produce, but also for slaughter-houses and purchase-associations for supplies of implements, fertilizers, and seed. Danish peasant farming defeated the large-scale manorial estates commercially through the officially encouraged efficiency of co-operative institutions. The Norwegian peasantry also developed co-operative institutions for purchase and marketing during the commercial breakthrough. But they never played the same crucial role as was the case in Denmark, since large-scale agrarian landlordism here was negligible as an antagonist force. The Norwegian co-operatives were more exclusively barriers towards market control by commercial middlemen.

The co-operative movement seems to have been even less dominant in Sweden, where the land-owning peasant interests had perhaps been more preoccupied with the formal privileges for the landed nobility. The development of specific peasant institutions in defence against the commercial non-peasant world was then probably more easily omitted. This incomplete argument will be somewhat further outlined in a subsequent chapter.

Peasant organization and public policy

The mercantilization of agriculture was accompanied by a remarkable change in social organization. The patriarchal localism of the old peasant society was split into three generalized principles at the national level:

Firstly, the public policy of the nation-building upper classes changed

148 Ibid., p. 78. A total amount of 506 mill. kg milk was delivered to the Swedish dairies in 1890, compared to more than 842 mill. kg in 1900. The modest increase in number thus implied a concentration of size.

from support of rural self-sufficiency towards integration of peasant agriculture in the economic mobilization of productive resources. The new economic framework was defined by conceptions of 'modernization', 'rationalization' and 'efficiency' as responsibilities of the state.

Secondly, the corporate peasant communities were opened up by horizontal peasant co-operation on the regional and national level. The peasantry itself increasingly took interest in co-ordinated action as a class, and gradually came to dominate the established agricultural associations.[149]

Thirdly, the peasant world was gradually permeated by impersonal market relations when even crofters and rural dependents were turned into wage labour. Where the commercial labour system supplemented family farming, ideal-type capitalism spread to the countryside.

Without going into detail, we will have to take a somewhat closer look at each of these principles.

Public policy. Norway, Sweden and Denmark all had their 'royal societies' for the official promotion of agriculture. Norway got the *Kgl. Selskap for Norges Vel* (literally: Royal Society for the Welfare of Norway) in 1809, which concentrated exclusively on agriculture from 1829.[150] Sweden got the *Kgl. Landtbruksakademien* (Royal Academy of Agriculture) in 1813,[151] while Denmark had its *Kgl. danske Landhusholdningsselskab* (Royal Danish Society of Agriculture) from 1769. These national organizations were, partially, established as superstructures on local agricultural associations led by public officials and–in Denmark–big land-owners.[152] In their policies towards the peasantry during the first decades of the 19th century, they stimulated versatile farming and peasant self-sufficiency. Enclosure–where not accomplished–was on the agenda, and so was information about soil improvement, better implements and seeds. While free trade was triumphing over protectionism from the late 1850s to the 1880s, the public policy towards peasant agriculture concomitantly changed from idealization of self-sufficiency–a particularly outspoken bureaucratic ideology in Norway–to advocacy of international division of labour.[152b] The specialization on commercial livestock husbandry, as a response to changing terms of trade, was stimulated by agricultural policy. Local agricultural schools and institutes had been established in all three of the countries towards the middle of the century while a national agri-

149 In all three countries there had been established agricultural associations by civil servants and the landed upper class early in the 19th century.

150 Cf. Sigvald Hasund, *Det kgl. Selskap for Norges Vel*, II, Gjøvik 1941, and Stein Tveite, *Jord og gjerning*, Oslo 1959, op.cit.

151 Cf. Juhlin Dannfelt 1913, op.cit.

152 Cf. Skrubbeltrang 1938, op.cit., pp. 100 f.; Juhlin Dannfelt 1913, op.cit., pp. 20 ff.

152b See Stein Tveite, *Jord og gjerning*, Oslo 1959.

cultural college was founded in Norway in 1854 and in Denmark in 1858.[153] The first official travelling agronomists started their work in all three countries in the beginning of the 1850s, while the annual agricultural meetings–which were semi-officially arranged from the 1840s –included exhibitions of machinery and cattle breeding from the 1860s.[154] The Norwegian agricultural bureaucracy for the promotion of productivity and commercial farming expanded from the 1870s. As in Sweden, farm partition was defined as a major problem, since the relatively prosperous peasant farmers, with a certain minimum of investment capital, were those most likely to profit from education and technical instruction. The public strategy for agricultural development was thus characterized by 'betting on the strong'.[155] In none of the countries was the state any longer a remote, alien, and demanding force outside the peasant communities. It gradually became an ambitious and present mechanism directly relevant to the organization of everyday life. The peasantry became intrinsically connected to a wider societal order through the dual intrusion of public power and commercial markets. Both had of course existed before, but their range expanded vastly during the second half of the 19th century.

Peasant self-organization. The co-operative movement in the sphere of commodity circulation developed after the individualization of production. The co-operative associations involved economic coalitions between local peasant groups. They were thus socially horizontal–without control by superior outsiders–and basically oriented to well-defined single interest purposes.[156]

The marketing co-operative movement became, as mentioned above, most developed in Denmark. The agricultural associations in all three countries were, however, increasingly characterized by peasant participation–to the relative displacement of officialdom and the rural upper class.[157] The local agricultural associations in Denmark were taken over by the new peasant proprietors parallel to the breakthrough of the co-operative movement.[158] The rural hegemony of the Swedish

[153] Cf. Ingers, op.cit., vol. III (Eskerød), p. 110; K. Hansen, ed., op.cit., vol. V, pp. 623 ff. and pp. 644 ff.; Dahl Jacobsen 1964, op.cit., p. 25 and pp. 30 f.; Borgedal 1966, op.cit., pp. 89 ff.

[154] K. Hansen, vol. V, op.cit., pp. 373 f.; Dahl Jacobsen 1964, ch. 2; Juhlin Dannfelt 1913, Appendix 20, p. LXXXV.

[155] Dahl Jacobsen 1964, op.cit., gives a careful examination of political system and public developmental strategy in Norwegian agriculture, while no real counterpart to this study exists for Sweden and Denmark. Cf. the Norwegian director of agriculture, Jonas Smitt, 'Om grundeiendommens udstrykning og dennes betydning for vort jordbruk', *Statsøkonomisk Tidsskrift,* Kristiania 1890, pp. 203–247.

[156] Cf. Eric Wolf, *Peasants,* 1966, op.cit., pp. 81 ff.

[157] The local *landbruksselskaper* (agricultural associations) as well as the general agricultural meetings, involved peasant participation after the middle of the century.

[158] Cf. Skrubbeltrang 1938, op.cit., p. 114.

landlords was–at least in the southern regions–only partially broken through co-ordinated non-noble and peasant action during the last decades of the 19th century. The potential unity of the peasantry broke down over the question of free trade, where the corn-producing sections demanded tariffs on their produce, while the deficit regions opposed it.[159] The corresponding conflict applies to Norway during the last decades of the century as well.[160]

The special purpose associations and the marketing or purchase co-operatives were built on individual production units–the single peasant farm. Inter-local and single-stranded relationships replaced the close primary group cohesion of the old peasant communities. The long-lasting vacuum left by the dissolution of the villages in Denmark at the end of the 18th century, was filled on a new contractual basis nearly a century later. The intermediate 'sack-of-potatoes' period was a protracted episode. The initial peasant freedom had been greater in Norway and Sweden, and the autonomy on the anonymous market thus probably implied a less dramatic change. But here too the traditional *gemeinschaft* character of peasant cohesion was transformed into greater functional specialization and secondary group relationships *before* the agricultural economy was left behind in a secondary position by industrialism.

Labour market. The paternalistic labour relationship dissolved in conjunction with the growth of commercial farming. The crofter system generally declined from about the middle of the 19th century (cf. chapter 6), and patriarchal labour control was replaced by more standardized legal contracts–with money as the mediating link between the opposite interests of employer and employee.[161] What did this labour market imply for the agrarian structure? We have seen that there was a notable rural population growth during the 19th century, which was not sufficiently absorbed either by farm partition or by urbanization–particularly not in Sweden and Norway. The individualized commercial agriculture at the same time meant an increasing socio-economic differentiation of the agrarian population, with a vastly expanding rural proletariate relative to the peasant proprietor class. This lower stratum of the rural population was generally pauperized during the 19th century, and became highly vulnerable on the impersonal

[159] Cf. Gellerman 1958, op.cit.

[160] Cf. *Det Norske Storting gjennom 150 år,* Gyldendal, Oslo 1964, vol II by Rolf Danielsen.

[161] Cf. Utterstrøm 1957, op.cit., I, pp. 825 ff. The commercially advanced manor farms in lowland districts of Sweden introduced the truck system–*statsystem*–already in the first half of the century, to secure a permanent and readily available labour force during modernization. The handicap, to the large land-owner, with copyholders and crofters was that they had to divide their working time between cultivating their own land and that of the manor farm. Cf. chapter 6 above.

labour market, since the closely-knit communities dissolved. Furthermore, their strategical position on this market deteriorated parallel to agricultural mechanization and change-over to livestock husbandry, which had been stimulated as a response to the shortage of farm labour. This development should be expected to produce an intolerable social tension, a rather explosive potential within the societal framework. But there was a safety-valve which prevented the pressure from boiling over.

Short note on emigration

The development of a landless rural proletariate became a heavier problem in Norway and Sweden than in Denmark with the population growth after 1815. Denmark had a greater proportion of urban dwellers–about double that of Norway and Sweden relative to total population–which eased off some of the rural population pressure. In Sweden the proportion of peasant farmers–including their families–relatively decreased from a half of the population in 1809 to less than a third in 1866, while the landless proletariate grew disproportionately. A corresponding downward mobility in the composition of the rural population also took place in Norway: while the number of peasant farmers (freeholders and tenants) increased by 50 % from 1801 to 1855, the rural landless more than doubled during the same period (cf. chapter 6). The general crisis was probably reinforced by the increasing farm partition, which had put a considerable number of farms at the margin of economic viability, and made them particularly vulnerable to the fluctuations of yields and market.

The mass emigration to North America became the major safety valve during the last three to four decades of the 19th century.[162] The aggregate comparative pattern is elucidated by the following figures:

Scandinavian emigration 1850–1910

	Total Number of emigrants	Emigration in percentage of excess of births
Norway	691 821	43.8
Sweden	966 662	31.4
Denmark	267 302	17.3

Source: Svalestuen 1971, p. 13.

[162] Cf. major works like Ingrid Semmingsen. *Veien mot Vest*, vol. I–II, Oslo 1941–1950; and Kristian Hvidt, *Flugten til Amerika*, Copenhagen 1971. A good survey is also presented in Hvidt, ed., *Emigrationen fra Norden indtil 1. Verdenskrig*, reports from a meeting of Nordic historians, Copenhagen 1971. A comparative outline is here presented by Andres A. Svalestuen, pp. 9–61.

While the Norwegian emigration–with the United States as nearly the only country of destination throughout the 19th century–was highest relative to the size of the population, the Danish counterpart definitely was least.[163] The wave of emigration is a rather accurate measurement of the economic cycles from the 1860s. The peaks occurred first in the years from 1866 to about 1873, then from 1879 to 1883, and finally in some years during the first decade of the 20th century.[164] Two-thirds of the total number of Norwegian emigrants until 1910 left the country before the end of the 19th century. The annual average was nearly 15 000 persons in the period 1866–1870. Then it fell to somewhat below 10 000 annually during the relative boom of the 1870s, after which it increased to well above 20 000 in annual average during the years 1881–1885.[165] More than 3/4 of the total excess of births was absorbed by emigration during the first half of the 1880s, with an absolute peak of nearly 30 000 emigrants in 1882–out of a population of less than 2 millions.

In all the Scandinavian countries, the emigration statistics were predominantly populated by labourers and servants.[166] In 1871–75 three-fourths of the Danish emigrants were classified as workers or domestic servants, while this applies to about half of the Norwegian ones. The Norwegian percentage of peasant farmers fluctuated between a tenth and a fourth of the total number of emigrants during the last three decades of the century–as in Sweden, while the peasantry more continually accounted for about a tenth of the Danish emigrants.[167] But we cannot therefore conclude that it was the poorest strata of the peasantry and of the proletarian landless that left. On the contrary, the most impoverished could probably not afford to escape during the early phase of emigration. The tickets were quite expensive, and generally a certain economic threshold had to be reached if emigration were to become a realistic alternative. The case seems rather to have been that the general impoverishment drove the somewhat less pauperized landless and smallholders to emigrate. The later reduction of travel prices, and the introduction of a system of pre-paid tickets, probably gave emigration a more distinctively proletarian character towards the turn of the century.[168] The traditionalistic character of backward, corporate communities also implied a barrier towards de-

163 On the European scale, the Norwegian emigration rate was outnumbered only by the Irish until the middle of the 1890s, when also Italy passed beyond the Norwegian rate.

164 Cf. Svalestuen 1971, op.cit., p. 12, and Ingers, op.cit., vol. III (Carlsson), pp. 320–331.

165 *Historical Statistics* 1968, Oslo, table 33, p. 57.

166 Cf. Svalestuen 1971, op.cit., table on p. 49.

167 Ibid.

168 Ibid., p. 17.

parture. Emigration was on the whole no collective or communalistic response to the strain caused by population pressure and scarce resources. It was an individualized or small group enterprise, a break with the traditional setting. The prerequisite for this surplus energy, above the cultural embeddedness, was a minimum of outward contact and geographical mobility. Those structural transformations which have been analysed in the preceding and the present chapter broke crucial barriers down, and became conditional for the exit option.[169] Emigration in turn stimulated the technical rationalization of agriculture, since the reservoir of cheap labour was drastically reduced.

'The Big Change'–towards a diagnosis

This chapter has been a rough outline of the transition towards neotechnical agriculture in its relation to the penetration of land and labour by the market system–commercially and financially. Our major ambition has been to emphasize the broad trend of the development at the level of the nation-state. We have touched on regional and local variations only where necessary for a rough survey of causal factors.

The process of change was–to all appearance–faster in making progress in Norway, although it started later, and from a technically more backward stage. But it should also be noted that many peripheral areas stayed outside the mainstream of agricultural mercantilization until well into the 20th century. The large-scale agriculture of Denmark and Sweden became market-integrated–based on neo-technical farming methods–relatively early: the Danish manorial farms probably during the latter part of the 18th century, and the Swedish ones on the whole a few decades later.[170] The principles of commercial agriculture then spread gradually to peasant farming, parallel to the enclosure movement in Sweden and the acquisition of peasant proprietorship in Denmark. In Norway, however, peasant farming initially dominated the rural scene, and the market-integration of the peasantry therefore implied a qualitatively new orientation of agriculture as a whole. The new principles for economic adaptation which gradually transformed the Norwegian peasantry were previously alien to the agricultural hinterland, since the rural aristocracy was negligible as a socio-economic linkage group. The commercially dynamic forces in the Norwegian countryside were rather to be found in fishing and forestry, which

[169] The pattern of mass emigration is of course an all too complex phenomenon to be thoroughly discussed here–cf. the standard accounts by Semmingsen 1941–50, op.cit., and Hvidt 1971, op.cit. See also detailed local studies like A. Svalestuen, *Tinns emigrasjonshistorie*, Oslo 1973.

[170] The development in the province of Scania, which had formerly been Danish land, was probably quite parallel to that in Denmark.

supplied parts of the peasantry with a supplementary income, although the major surplus was probably accumulated by legally favoured urban merchants.[171]

The outcome all over Scandinavia was an individualized farmer class with some co-operative market institutions for the circulation of the products of mixed farming–arable and livestock, with the growth of animal husbandry being crucial to the commercial breakthrough.

Literary historiography in Norway has dated this quite dramatic breakthrough to a few decades in the second half of the 19th century.[172] The transformation hardly occurred with such dramatic suddenness as advocated in the more extreme interpretations, although a notable change definitely occurred within a relatively short period of time, as indicated above. We have seen, however, that substantial innovations in agricultural techniques, patterns of trade, and credit supply were already spreading gradually during the first half of the 19th century, even in Norway. But the regional differences were substantial, and the pattern of variation between local types was quite vast. The transformation could take place rather quickly in certain communities and rural districts, when opened up by main roads and country stores.[173] The overall national pattern is therefore contingent upon two different developmental sequences: firstly, the scope and speed of each single local transformation; and secondly, the degree of diachronous correspondence between the various local patterns. A substantial amount of detailed and co-ordinated research would be needed to give only an approximately definite statement of this intricate mixture of synchronous and diachronous complexities. Our present ambitions have necessarily had to be far more modest.[174]

The problem of the agricultural *hamskifte,* the 'big change', is particularly accentuated in Norway, since no large-scale and market-oriented farming had paved the way as in Sweden and Denmark. But the interpretations of the characteristics of the big change are controversial. One well-established theory says that the change implied a definite qualitative transformation from relative self-sufficiency to pro-

171 Cf. chapter 5 above. Some of these acquired land themselves, and became 'bourgeois proprietors'. A few reinvested their mercantile surplus to make 'model farms'.

172 The standard reference here is Inge Krokann, 'Det store hamskiftet i bondesamfunnet', *Norsk kulturhistorie*, vol. V, Cappelen, Oslo, 1942, pp. 100–194. The detailed examples in Krokann's general account are mainly from the commune of *Oppdal* in the mountainous central east. The big *hamskifte* in the peasant society has been a central theme in Norwegian *belles-lettres*, and 'the big change' is a rather poor translation.

173 The native district of Krokann–1942, op.cit.–gives probably an example of a rather sudden transformation.

174 Cf. the source evaluation in a crucial article by Andreas Holmsen, 'Kulturskiftet i Noreg kring midten av 19. hundradaaret', in *Gard, Bygd, Rike*, 1966, op.cit., and also the review article by Kåre Lunden in *Historisk Tidsskrift*, vol. 49, op.cit.

duction for the market, from subsistence farming to commercial farming.[175] This view implies that the Norwegian society was fundamentally dualistic until about the middle of the 19th century, with the non-capitalist rural world of subsistence-oriented peasant communities on the one hand, and the commercial urban world of money, profits, trade, and markets on the other. These two societies were, presumably, related to each other, as Centre to Periphery.

The opposition to this dual society thesis–and indeed to the very notion of the *hamskifte*–argues that the picture of self-sufficient peasant communities prior to the presumed 'big change' is false, and that the peasantry were always guided by the most profitable kind of economic activity.[176] This interpretation has been supported by reference to the regional division of labour between arable and livestock areas, to the persistent need for a certain amount of cash for taxes, rents and non-agricultural requirements, and to the indisputable fact of widespread peasant trade within the old society.[177]

The dualistic society thesis was, in our view, right in so far as there occurred a notable qualitative change in peasant economy and society with the transition towards individualized neo-technical farming and greater dependence upon the anonymous market.

The counter-view is right in so far as peasant trade and regional division of labour took place from very early on, and so far as the market was already then a strategic factor subject to speculation or withdrawal according to economic circumstances.[178]

This answer indicates that the problem definition as stated could be somewhat misleading.

We have to distinguish clearly between 'self-sufficiency' and 'subsistence-orientation'. The old peasant society was of course never self-sufficient, but peasant trade–as well as farming and subsidiary occupations–did nevertheless aim at the fufilment of the consumption needs of the family (and claims by outsiders) rather than at profits for reinvestment. The demand for additional capital was primarily restricted to the releasing of hereditary shares, until the technical improvement of the new farming opened up another agricultural dimension–the

[175] Cf. Steen 1957, op.cit., who gives the most distinguished expression of this interpretation.

[176] Cf. Tveite 1959, op.cit., pp. 31–72.

[177] Ibid., pp. 71 f. A somewhat critical but nevertheless quite balanced review of Tveite's work is given by Andreas Ropeid in *Historisk Tidsskrift*, vol. 40, no. 2, 1961, pp. 382–388. Ropeid argues that the undisputable quantitative development led to a qualitatively transformed type of economic adaptation, and he thereby, like Steen, supports the notion of the *hamskifte*.

[178] Cf. chapter 1 above. Cf. also the fact that there seems to have occurred a notable withdrawal back to greater self-sufficiency with the crisis during the first three decades of the 19th century. Tveite, op.cit., is definitely right in stressing this fact.

good return on long-range investment. Peasant proprietorship was a basic prerequisite for the new interest in this. But not even the traditional subsistence economy was inconsistent with a certain dependence upon trade and markets, as discussed theoretically in chapter 1. The market in a peasant economy may be regarded as a strategic option, which is employed so far as required by the balance of livelihood. But peasant trade is then oriented towards the subsistence requirements of the family household, and it is principally performed as a seasonal barter activity. This element of parochial reciprocity should be dissociated from the anonymous institution of price-regulating market exchange, although there always is a question of principal emphasis rather than pure types.[179]

The agricultural revolution implied that the peasant farm gradually gained some more pronounced characteristics of the capitalist business enterprise. The reciprocal principles of communal work unmediated by money–like the *dugnad* institution (cf. chapter 5)–declined, while the peasantry developed into professionalized agriculturalists. But important elements of non-mercantile domain survived in spite of growing market exchange. Particularly in Norway, land continued to a great extent to be disposed of by kinship domain through the allodial law, and the disattachment from the soil never went so far as to make farming a pure investment option for the agricultural population. It would then be misleading to label the transformed peasantry a capitalist proprietor class in the Marxian sense, although some elements of this character definitely were intrinsic to the transformation. The label is more appropriately attached to the bourgeois proprietors who bought land with a surplus gained through other types of trade and industry, as well as to the big land-owners of Denmark and southern Sweden, who exclusively based their farming upon hired labour.

The dualistic character of the old society definitely became weakened during the agricultural revolution–and the institutionalized dichotomy between centre and periphery, urban and rural, became blurred. This was principally due to a dual development:

Firstly, to the intrusion of the money and market economy, which integrated the peasantry into the wider national order and strengthened its horizontal class characteristics across the corporate communities.

Secondly, to the concomitant differentiation and specialization of the rural population, which implied a growing rural proletariate corresponding to the urban plebs.

This dual development thus pointed towards a general class consoli-

179 The classification of socially integrative mechanisms employed by Karl Polanyi and his school–cf. chapter 2 above–seems fruitful here.
Cf. my own qualification of these perspectives–as applied to the Norwegian *hamskifte*–in 'Nytt perspektiv på det store hamskifte', *Historisk Tidsskrift*, Nr. 2, 1975.

dation—at the bottom of society as well as at the top—across the former antagonism between rural agrarianism and urban interests—particularly affecting those areas where the socio-economic differentiation of the peasantry was most pronounced.

The effects of the agricultural revolution in Scandinavia should then be commented on in relation to those socio-economic dimensions of dependence/autonomy which were discussed in chapter 4.[180]

The big change had an equivocal consequence for the *security of land occupancy*. It was formally increased for those substantial parts of the peasantry that obtained freehold ownership. Danish landholding was particularly transformed in this respect during the 19th century, but it also affected remaining leaseholders in Sweden and northern Norway in the second half of the century. Landholding security at the same time decreased for those outfield squatters and other marginal groups who had made a living from the commons. But the intrusion of market forces also implied greater vulnerability even for the peasant proprietors. The mortgage debt to financial institutions increased the rate of bankruptcy and distraint, while increasing farm partition—particularly in Norway and Sweden—made the dependence on working capital for cash crop production a precarious living. This is partially reflected in the emigration statistics, as outlined above.

The *degree of managerial responsibility* generally increased, so far as enclosure, new methods of farming, and monetary economy implied the dissolution of communal landholding and a notable individualization of agricultural adaptation. This development admitted greater managerial flexibility, where the individual farmer became free to choose any crop rotation or any mixture of arable and livestock he pleased. The individual agriculturalist became, instead, directly subject to anonymous market forces, without mediation from overlord or community.

The effects on the *relationships to the vertical market* are again somewhat ambiguous. Owing to the initial dependence of the Danish peasantry, the unmediated access to circulation of produce here definitely increased—particularly when this autonomy became collectively structured by marketing co-operatives. But the original freeholders of Sweden and Norway may have had some of their mobility in trade restrained by the new class of professionalized commercial middlemen. The new network of country stores was less open to commercial bargaining than the trade institutions of the old barter economy. It is likely, however, that this is only a minor counter-trend to the commercially liberating effects implicit in the abolition of mercantilistic privileges around the middle of the century.

The *family and community orientation* was definitely transformed.

[180] The political status of peasantry will be discussed in the next chapter.

The individualization of farming also implied a concentration around the nuclear family, even if the three-generation household survived. The community fellowship was increasingly replaced by secondary groups like special-purpose associations, and institutions of a horizontal and single-stranded kind. The factional inclusion of labourers and crofters within the main household was replaced by more contractual, and less embracingly patriarchal, relationships.

In general terms, we might conclude that the introduction of neotechnical agriculture implied greater stability, predictability and individualized peasant control in the sphere of production, while peasant insecurity was transferred to the circulation of credit and produce–the market as an opaque sphere of aggregate exchange.

PART III

Political framework and peasant voice

We are now, in Part Three, going to trace the development of peasant action within the Scandinavian macro-political setting. The account will, by necessity, have to be rather conventionalized, since minute details and local variations should not be allowed to put the basic comparative characteristics into the shade. By 'politics', moreover, we are not concerned with that type of micro-politics which is confined to the local horizon of the peasant community. The dual intrusion of market relations and state power integrates the peasantry within the wider national order, and we are concerned with the concomitant political articulation within this order, in relation to other social classes and to more comprehensive institutions at the level of the nation-state.

CHAPTER 8

Peasantry and political system

The integration of the peasantry within the political system has three crucial aspects which will be comparatively outlined in the present chapter (cf. chapter 4 on the dimensions of peasant dependence and autonomy). Firstly, the status of peasantry within the polity–i.e. the development of peasant citizenship and legitimate participation in public affairs; secondly, the actual peasant employment of their legitimate rights in national politics; and thirdly, the organization of local government within the national polity, and peasant participation at this level. This account thus serves as a necessary introduction to the qualitative substance of peasant politics, which is presented in the subsequent chapter.

The systemic context

The type of national integration–by constitution and regime–lays down the rules of the game, and defines the framework of political action, whether this is accepted as legitimate or made subject to challenge from below. The peasant politics are modelled within a structure of institutional constraints, and we will roughly consider three aspects of this interrelationship:

Firstly, the Scandinavian peasantry face different types of state formation, styles of rule, and constitutional arrangements. The character of the *regime* will have to be accounted for.

Secondly, the peasantry are faced with formal thresholds of political participation. The development of their political status during the 19th century is expressed by a sequence of barriers which are passed within the polity.

Thirdly, the peasantry are gradually more firmly integrated into the national political system, parallel to the development of new types of authority relations between the peasantry and the wider social order.

The regime[1]

The twin kingdom of Denmark–Norway was–as we saw in chapter 5–a monarchic-absolutist regime from the *coup d'état* of 1660. The constitution was made formally absolutist by the Royal Law (*Kongeloven*) of 1665, implying–in principle–more unrestricted power to the King than any other absolutist constitution in Europe at the time.[2] But in fact the regime was quite benevolent, liberal and progressive–probably more so than those of most other European countries. The bourgeoisie increasingly took over the state apparatus after the power of the old feudal aristocracy had been broken by royal force, and the Danish peasantry were gradually liberated from above during the last decades of the 18th century (cf. chapters 5 and 6). Royal absolutism was increasingly legitimated by a theory of accountability to public opinion, which paved the way for constitutional reform.[3] *Norway's* new constitution after the separation from Denmark in 1814 was based on the principle of separation of powers between the king and his cabinet, on the one hand, and the representative assembly, the *Storting*, on the other.[4] The constitutional history for the next fifty years was dominated by the precarious balance between these three organs of power.[5] At first–until the 1830s–there was a permanent political tension between the Swedish-Norwegian king–with the cabinet as a royal tool–and a *Storting* dominated by public officials. The king was then gradually pushed into the background–while the civil servants controlled both these organs. This dual bureaucratic power was gradually challenged as the civil servants lost some of their dominant position in the *Storting*,

[1] See inter alia, Nils Andrén, *Government and Politics in the Nordic Countries,* Almqvist & Wiksell, Stockholm 1964; Andrén, *Svensk Statskunskap,* Liber, Stockholm 1961; Nils Herlitz, *Grunddragen i det svenska statsskickets historia,* Svenska bokförlaget, 6. ed. Stockholm 1964; Alf Ross, *Dansk Statforfatningsret,* 2. rev. ed., Nyt Nordisk Forlag, Copenhagen 1966; Torstein Eckhoff, *Utviklingslinjer i norsk statsforfatning,* Universitetsforlaget, Oslo.

[2] Cf. Ross 1966, op.cit., pp. 86 ff.

[3] See Jens Arup Seip, *Teorien om det opinionsstyrte enevelde,* Universitetsforlaget, Oslo 1958.

[4] See, inter alia, Steen, *Det frie Norge,* vol. I, "1814", and Mykland (ed.), *Omkring 1814,* Oslo 1967.
The *Storting* was to impose taxes, authorize expenditures, pass laws, and control the administration, while the King appointed his cabinet without any interference from the representative assembly. The ministers were completely independent of the lagislators, and could not take part in the deliberations of the *Storting*. The King had a right to refuse his sanction of legislative decisions, but only with a postponing force, since a decision became valid law without royal consent if adopted by three successively elected assemblies. There was a special Norwegian cabinet in addition to the Swedish one.

[5] Cf. Eckhoff, op.cit., and Kaartvedt, *Det Norske Storting gjennom 150 år,* vol. I, Gyldendal, Oslo 1964.

while they consolidated their dominion of the cabinet and increased their influence over the gradually less powerful king. The emerging tension between a government of public officials and a *Storting* where their colleagues now were in a minority–while the majority made up of peasants and pro-liberal middle class representatives constituted the opposition–led to the introduction of parliamentary democracy from 1884. The final constitutional struggle in that year broke the peculiar political system of the 'civil service state', and made the government responsible to the majority in the representative assembly.[6]

The peculiarity of Norway after 1814 is thus that the bureaucracy was the basic nation-building elite, and not the nobility (even noble titles were formally abolished in 1821), nor even the bourgeoisie. The bureaucracy pretended to represent the general national interest against the sectional one of the advancing peasantry, and built their position on a theoretical and practical identification of 'politics' with 'administration'.

In *Denmark* too the absolute monarchy became in fact a rule by bureaucracy, as the role of the civil servants became gradually more important throughout the 18th century. Constitutionally, royal absolutism prevailed unrestricted until the beginning of the 1830s, when the decision was made to establish four consultative provincial assemblies.[7] Constitutional monarchy–as an intermediate type of regime between royal absolutism and representative democracy–was a political trend in Europe after the revolution of July, and the Danish provincial assemblies were a substitute for such a regime.[8] The recommendations of the assemblies were only consultative–even if they often must have been difficult to ignore.[9] The system marked a transitional period whereby the people–i.e. the peasant proprietors and the liberal bourgeoisie–were prepared for representative government under the guardianship of the State. The provincial assemblies effected the transition from royal absolutism to a liberal constitution.

The Constitution of June (*Junigrundloven*) from 1849 recognized–as a result of popular pressure–a division of power, with legislative and

[6] Cf. Alf Kaartvedt, *Kampen mot parlamentarisme 1880-1884,* Univ.forl., Oslo 1956, and J. A. Seip, *Et regime foran undergangen,* Gyldendal, Oslo 1945.

[7] These *rådgivende provinsialstænderforsamlinger* were not a result of popular pressure, but they were initiated from above. The Danish king had been forced to grant such privileges to the duchies of Slesvig and Holstein by a separate constitution, and the system was extended to the Kingdom–with one Assembly of Estates for the islands (in Roskilde) and one for Jutland north of Slesvig (in Viborg). The decision was made in 1831, with a supplementary decree of 1834, and the assemblies met for their first session in 1835–36. Cf. Hans Jensen. *De Danske Stænderforsamlingers Historie,* vol. I–II, Cph. 1931–34.

[8] Cf. Ross 1966, op.cit., pp. 98 ff.

[9] Cf. Jensen, vol. I, 1931, op.cit., pp. 543 ff.

judicial controls on the king and his bureaucracy. The legislative power vested partly in the king, and partly in the two-chamber national assembly, the *Rigsdag,* whose consent was necessary when the government levied taxes.[10] The two chambers, the *Folketing* and the *Landsting,* were equal in power, but they differed in composition, since the representatives to the latter house were subject to certain economic qualifications, as we shall see below.

This liberal constitution suffered a notable backlash in a revision of 1866, when the conservative forces, led by the higher bureaucracy and the great landowners, took advantage of the national defeat over Slesvig-Holstein, when Bismarck annexed the duchies. The anti-German policy on the duchies had been a *raison d'être* for the bourgeois national-liberals, while the aristocracy was less committed on this issue. The national-liberal policy for a Danish Slesvig had led to the defeat, and the now dominant conservative party took over the upper chamber in the *rigsdag* by reserving exclusive representation to the wealthier classes, in addition to the right to appoint 12 of the 66 *landsting* members given to the king.[11] Parliamentary democracy–the principle of government responsibility to the majority in the lower chamber–was finally secured with the 'change of system' (*systemskiftet*) in 1901, after thirty years dominated by a constitutional struggle between the conservatives and a peasant-based liberal majority in the *Folketing.*[12]

Absolute monarchy in *Sweden* ended, as we saw in chapter 5, with the death of King Charles XII in 1718. The noble landholding aristocracy now became dominant, partly by a monopoly of the higher appointments in public office, and partly through its position in the four-tier Assembly of Estates, the *Riksdag,* which was founded on a specific type of representational elitism emanating from the nobility, the clergy, the bourgeoisie, and the peasantry. The Freedom Era (*Frihetstiden*) from 1717 to 1772–called after the power of the Assembly of Estates relative to the King–meant a rather meagre influence from the unprivileged lower Estates, until they became instrumental to the reassertion of royal power during the reign of Gustav III and Gustav IV Adolph from the last decades of the 18th century. The aristocracy, which at first supported the monarchic *coup d'état* in 1772, owing to increasing opposition from the lower Estates in the *Riksdag,* soon revived its constitutional opposition, and became politically over-

[10] See *Den Danske Rigsdag 1849–1949,* vol. I, Schultz forlag, Copenhagen 1949, pp. 41–94, with the *Junigrundlov* as an appendix on pp. 343–359. Cf. also Kenneth E. Miller, *Government and Politics in Denmark,* Houghton Mifflin, Boston 1968, p. 33.

[11] Cf. *Den Danske Rigsdag,* vol. II, pp. 5–36, with the revised constitution of 1866 as an appendix on pp. 709–722.

[12] Cf., inter alia, Ross 1966, op.cit., pp. 101–126; Povl Engelstoft in *Den Danske Rigsdag,* vol. II, op.cit., pp. 5–280; Miller 1968, pp. 34 ff.

shadowed through the reintroduction of absolutism from 1789.[13] The Gustavian era ended when Gustav IV Adolph was dethroned in 1809, and the new constitution was drafted to re-establish the separation of powers between the king and the four-tier Assembly of Estates.[14] The Swedish constitution of 1809 established a separation of executive, legislative, and judicial powers, and in this respect it adopted the basic principle of the Norwegian constitution of 1814 and the Danish one of 1849. But unlike Norway and Denmark, the representation was still exclusively based on legally distinct Estates, which were nothing like organized social groups. The socio-economic development of the agrarian society, which was outlined in the preceding chapters, made these Estates increasingly more problematic as criteria of political division and articulation of interest, as we shall see below. The four-tier Estates Assembly was then replaced by a two-chamber *Riksdag* in 1866, where economic position–by rather strict qualifications for suffrage and eligibility to both chambers–was substituted for the outdated Estate connection. Parliamentary democracy was not introduced until 1917–thirty-three years later than Norway and sixteen years later than Denmark.[15]

Thresholds of political participation

The rough outline above is only a starting point: it says nothing about that political integration of peasantry which actually took place during the 19th century. We are interested in the internal comprehensiveness of the nation-state, the extension of the scope of the polity which lay implicit in the constitutional development. To make a complex historical evolution immediately comprehensible, we will employ an analytical scheme developed by Stein *Rokkan* in his studies of the extension of citizenship as a prerequisite for the development of modern party systems.[16] The sequence of thresholds passed in a transition from sub-

[13] Cf., inter alia, Sten Carlsson & Jerker Rosén, *Svensk Historia,* vol. II, Bonniers, Stockholm 1964.

[14] The oppositional *coup d'état* was based on the aristocracy, particularly the higher circles of the civilian and military officialdom, and the deposition of the King was approved by the Assembly of Estates. Cf. Herlitz 1964, op.cit., pp. 190 ff.; and Andrén 1961, op.cit., pp. 18 ff.

[15] Andrén 1961, op.cit., pp. 25 f.

[16] See Rokkan, *Citizens, Elections, Parties,* Univ. forl., Oslo 1970, pp. 72–144, pp. 226–247; and S. M. Lipset and S. Rokkan (eds.), *Party Systems and Voter Alignments, The Free Press, New York 1967, intorduction, pp. 1–64.* The scheme is fruitfully employed by W. M. Lafferty, *Economic Development and the Response of Labor in Scandinavia,* Univ.forl., Oslo 1971, pp. 115 ff.

ject status to gradually more advanced levels of formal participation is then conceptualized as follows:[17]

- the *legitimation* threshold concerns the question whether all protests are officially rejected as conspiratorial, or whether there is some recognition of the right of petition, criticism, and opposition.
- the *incorporation* threshold affects the degree to which all or most of the supporters of a certain movement are denied status as participants in the choice of representatives, or whether they are given political citizenship on a par with their opponents.
- the *representation* threshold concerns the conditions under which new social movements initially must join larger and older movements to ensure access to representative organs, or whether they can gain representation on their own.

When trying to specify how the peasantry were affected by this sequence of barriers, it should be noted that the scheme does not intend to catch the implication of socio-economic structure as a mechanism of control. That perspective–which hinges upon the analysis of chapters 5–7 above–will be left for the subsequent chapter, while the *actual* political participation of the peasantry will be discussed in later sections of the present one.

The *legitimation* threshold–the barrier against legitimate opposition from the peasantry–is impossible to specify by clear-cut quantitative indices. It concerns the official recognition of a right of voice from below. The barrier was passed not only gradually, but in a quite irregular development, with temporary setbacks during instances of acute peasant mobilization, even if the long-term trend all over Scandinavia was a growing compliance.[18]

In Norway there was a very old institutionalized tradition of petitions to the king.[19] The king was seen as the incarnation of established legal rights, and the peasantry complained about undue taxation, about injustice from the local officialdom, or about commercial exploitation from the licensed merchants. This practice of direct appeal to the king

17 The last stage–when executive majority power is established–is not included in our discussion, since it generally occurred later then the period we are considering. The principle of numerical majority rule is estimated by an analysis of the built-in checks and counter-forces against it.
We employ the scheme, of course, without any evolutionist conceptions beyond the descriptive purpose.

18 To account systematically for legal rights, and freedom of speech, expression and assembly, during a century in three countries, would, of course, be an overwhelming task. We only intend to indicate some very rough characteristics.

19 See the numerous examples from the Middle Ages to the end of the 18th century analysed by Koht, *Norsk Bondereising,* Aschehoug, Oslo 1926.

implied a mutual legitimation of the regime and the subject peasantry. It established the right of voice from below, and it confirmed the position of the ultimate power and the style of rule of royal absolutism. This particular channel of protest thus also meant a diversion from structural opposition. In the 1780s, though–after several peasant petitions to the king–the bureaucracy obtained a general rejection of mass complaints: the person who *wrote* the letter should share responsibility for its contents.[20] It has, moreover, been thoroughly documented how local or provincial peasant movements beyond the legitimate type were repressed both before and after 1814.[21] The leaders were rather severely prosecuted after provincial peasant uprisings in a western area in 1765, and in the southern districts in the middle of the 1780s.[22] When peasantry from some eastern provinces marched against the capital in 1818, to protest against the new *Storting* and its imposition of extraordinary taxes, the public reaction was substantially milder, although the army was sent out even on this occasion.[23] The prosecution of the leader of the rural proletarian movement around the middle of the 19th century was quite mild relative to the fate which hit the rebellious peasant leader Christian *Lofthus* half a century earlier.[24] We shall have a closer look at these types of peasant unrest in the subsequent chapter, so it suffices here to ascertain that even if popular agitation and 'subversive activities' were forbidden, and in principle remained so during the period we are analysing, there definitely occurred some liberalization both in legal practices against intolerable opposition, and in the delimitation of those types of political expression which were considered intolerable. [25]

This account is not unreservedly relevant to Denmark, because the Danish peasantry were directly subject to the jurisdiction of local landlords until the late 18th century (cf. chapter 5). The strong paternalistic tradition gave little scope for tactical mobility from the peasantry, since the relevant figure of appeal was the lord who was immediately dominant on local lands, rather than an elevated incarnation of supralocal law. When the repercussions of local jurisdiction disappeared throughout the 19th century, the peasantry also got a socio-economic

[20] Cf. Steen, *NFLH,* vol. VII, op.cit., pp. 109 f.

[21] See Koht 1926, op.cit., and Steen, *Det frie Norge,* vol. II, Cappelen, Oslo 1953.

[22] See chapter 9 below, and notes 4–12 to that chapter.

[23] See chapter 9 below, notes 20–26.

[24] See chapter 9 below, notes 38–43.

[25] Cf. J. A. Seip's analysis of the development of freedom of speech and expression towards the end of monarchic absolutism–Seip 1958, op.cit.
The constitution of 1814 said nothing about freedom af assembly, although freedom of speech eas explicitly established. On the notable evolution of different voluntary associations from the 1830s, see Steen, "De frivillige sammenslutninger og det norske demokrati", in *Tusen års norsk historie,* Cappelen, Oslo 1958, pp. 182 ff.

foothold to act on in the political arena. When the popular movement became politically visible in the beginning of the 1840s, the government promptly answered with a prohibition against freedom of assembly in the rural areas.[26] But this act was soon repealed, and the peasantry were granted all formal rights of political participation shortly after.[27] Both in Norway and Denmark the most notable legitimation of a politicalized peasantry is thus expressed by the constitutional reforms of the 19th century.

The inherited representation of Estates made the situation in Sweden rather different. Freedom of assembly and association had been rather restricted under the *Vasa* kings before 1780, but a notable liberalization occurred towards the end of the Freedom Era. The Gustavian restoration meant some tightening of liberal regulations, while they were loosened anew after the reformation of 1809: The provisions against freedom of assembly and association were liberalized by a decree of 1816, and remaining restrictions were removed in 1849.[28] But these regulations had not been directed against the political articulation of peasantry as such. The Estate constitution (*stenderforfatningen*) granted formal representation to the peasant population, and already formed a legitimate channel for peasant politics during the 18th century. The Swedish peasantry were, however, split by the strict regulations of the boundary around the peasant Estate, which included mainly freeholders and holders of crown land (*skattebönder* and *kronobönder*), while the dependent peasantry on noble lands (the *frälsebönder*) were excluded.[29] Peasant mobilization outside the legitimate representative channel was repressed. When the peasantry of Dalecarlia refused to perform military service during the war against Russia, and marched to Stockholm in 1743, they were beaten down by armed force, and the leaders were prosecuted while fidelity was demanded from the common people.[30] The unrest among the *frälsebönder* of the southern provinces of Scania and Halland during the 1770s–directed against the heavy duties on manorial estates–was also suppressed with public support, and without any intervention from the relatively privileged peasant Estate of the *Riksdag*.[31] The legitimate peasant articulation

26 See chapter 9 below, notes 69–71 to that chapter.

27 By the Constitution of June–*Junigrundloven*–1849.

28 Cf. Gunnar Wallin, *Valrörelser och valresultat, Andrakammarvalen i Sverige 1866–1884*, Sth. 1961, p. 334–with further references.

29 Cf. Ragnar Olsson, *Bondeståndet under den tidigare frihetstiden. Val, organisation, och arbetssätt*, Lund 1926, pp. 26 ff. The peasant owners of *frälseskattehemman* were also included, and so were the conscript peasantry on military 'fiefs', *forläningsbönderna*.

30 See chapter 9 below, notes 90–91.

31 *Cf.* Staffan Smedberg, *Frälsebonderörelser i Halland och Skåne 1772–76*, Läromedelsförlagen, Uppsala 1972.

under the Estate constitution originally implied that peasant complaints, *besvär*, from one or two counties should be gathered by the chief local official, the *heradshövding*, and put forward through the *Riksdag*.[32] Organized peasant dissatisfaction was thus effectively prevented through institutional devices, which were reinforced by the co-optation of the peasant Estate. The representatives were raised from their local setting to participate–without much influence–in central decisions, which later made them responsible within their own community.[33] 'Legitimation' might thus be a double-faced complicance. The early co-optation of the Swedish peasantry might partly explain the lack of anti-system mobilization among the *skatte-* and *kronobönder*, while the dependent *frälsebönder*, who were outside this representative system, remained marginal to the Swedish agrarian structure, and thus were unable to trigger off any substantial destabilizing movement (cf. chapter 9 below). Given the reservations which will be discussed subsequently, this system formally prevailed during the first two-thirds of the 19th century, although it was gradually undermined by the socio-economic transformation of Swedish agrarian life.

The *incorporation* of the peasantry, with an explicit status within the political system, implies a step further than merely granting them a legitimate right of voice. The framing of franchise rules can be used as a direct indicator of the degree of incorporation and political citizenship.

The Norwegian constitution of 1814 virtually based the regulation of suffrage on a compromise between three general criteria for limitations of political citizenship:[34]

- the estate principle gave the vote to male heads of household from the peasantry and the *burghers.*
- the *censitaire* principle implied restrictions based upon real property, or upon minimum tax assessments on incomes.
- the competence principle gave the vote to public officials by virtue of their profession.

The original provision stated that suffrage for men over 25 years of age should be given to all the king's officials, to citizens who were licensed as merchants or artisans, or who owned urban property of a fixed value, and to those of the peasantry who were freeholders or who

[32] Cf. R. Olsson 1926, op.cit., ch. III, pp. 119–147.

[33] Here the decisions about taxes were particularly important, – cf. Olsson 1926, op.cit., pp. 161 f.

[34] Cf. the general discussion of suffrage criteria by Stein Rokkan and Reinhard Bendix, "The Extension of Citizenship to the lower Classes", in Mattei Dogan & Richard Rose (eds.), *European Politics,* Little Brown, Boston 1971, p. 18.

had leased registered land for more than five years.[35] The three basic criteria thus disqualified the urban proletariate and the rural landless from the vote. The franchise regulations after 1814 also required that those qualified had to register, by taking an oath on the Constitution, to be allowed to participate in the elections (which were indirect during the 19th century). It has been estimated that about 10 % of the population qualified for the vote in 1814, while about 7 % were registered. This proportion of 10 % qualified, which is influenced by the demographic development, the age composition and the male/female ratio, had dropped to 7.6 % in 1859 and to 7.4 % in 1870, while it rose to 9.4 % after the new economic qualifications were established in 1884.[36] But the proportion of enfranchised relative to the total population does not tell how discriminating the regulations were on social class. During the period investigated here, this is more properly indicated by the number of qualified relative to all men above the minimum age. From 1814 to 1859 we know the number of registered voters only. Two methods have been employed to estimate the number of those who were qualified. When we know the ratio of registered to qualified voters from 1859 onwards, a simple retropolation indicates that about 45 % of all men above 25 years of age were entitled to vote in 1814.[37] The result of this method has recently been compared with an estimate based on census data and other available information on the number of farms and farmers, related to the constitutional specifications for qualified suffrage.[38] The outcome is strikingly identical to the one based on retropolation of the first known ratio, so the estimate that about 45 % of all men above 25 were qualified for the vote in 1814 is probably fairly valid. Provided the ratio from 1859 is approximately correct for other years too, the proportion of enfranchised male voters had dropped to 39 % in 1825, and further to 35 % in 1845, while it was 33 % in 1865, and 34 % in 1875, until it rose again after the new economic qualifications of the 1880s.[39] This development was due to the social development towards the middle of the 19th century (cf. chapter 6), rather than to reversals in the constitutional regulations. In a comparative perspective–both in a European and a Scandinavian context–the Norwegian constitution of 1814 was extraordinarily radical and democratic, particularly as concerns the political inclusion of

[35] Article 50 of the original constitution. Cf. *Det Norske Storting gjennom 150 år,* vol. I, op.cit., pp. 51–68.

[36] Ibid., vol. I, p. 52.

[37] Cf. Rokkan, "Norway: Geography, Religion and Social Class", in Lipset & Rokkan 1967, op.cit., p. 380.

[38] See Stein Kuhnle, "Stemmeretten i 1814", *Historisk Tidsskrift,* nr. 4/1972, pp. 373–390.

[39] Cf. Rokkan in Lipset & Rokkan 1967, op.cit., p. 380.

the peasantry. Constitutional protections against peasant power were established in regulations of representation and constituencies, as we shall see; but the potential of peasant influence was quite large. The precarious international situation connected to the dissolution of the Danish-Norwegian union might have conditioned a strong need for national solidarity and legitimation of elite policy from below, especially since the traditionally independent Norwegian peasantry also constituted the backbone of national military defence.[40] The democratic constitution was then quite vigorously defended by the Norwegian bureaucratic elite during the decades immediately after 1814, since it meant the main national protection against Swedish pressure for 'amalgamation'–which would have been to the obvious benefit of Sweden as the dominant partner–within the new union.[41]

In the Provincial Estates of Denmark (1831–1849), franchise was given to men with landed property above a certain minimum, and to leaseholders on the greater farms, apart from the officials and the *burghers*. This implied that peasant smallholders were excluded even if they possessed proprietary rights to their land. It is difficult to estimate how discriminating these regulations actually were, but it should also be noted that their social bias was reinforced by the fact that freeholders and greater leaseholders constituted a smaller proportion of the rural population then the corresponding groups in Norway (cf. chapter 6). We know that slightly less than 3 % of the population in the Kingdom were entitled to vote under these provisions, and this might–carefully–be compared to the 10 % of the total population which qualified for suffrage in Norway in 1814.[42]

The two-chamber *Rigsdag* of 1849 implied a general enfranchisement of all men above 30 years of age. There were, however, some minor qualifications, which excluded paupers, together with servants and farm helpers without their own household. It is thus estimated that ca. 27 % of all men were excluded for reasons other than age.[43] The conservative reversal for the upper house in 1866 introduced a weighted tax vote for elections to this chamber.[44] Generally, 14–15 % of the population were allowed to vote in the first decades after 1949–

[40] Cf. Koht, "Trongen til demokrati 1814", *Historisk Tidsskrift,* vol. 34, 1947, pp. 133–151. See generally Mykland (ed.), *Omkring 1814,* op.cit., where Koht's article also is reprinted. See further Steen, *Det frie Norge,* vol. I, op.cit. on the events in 1814.

[41] Cf., inter alia, Steen, "Bøndene kommer", in *Tusen års norsk historie,* op.cit., pp. 177–181.

[42] Cf. Kjeld Winding, "Danmark", in *Framveksten av de politiske partier i de nordiske land på 1800-tallet. Problemer i nordisk historieforskning II,* Universitetsforlaget, Bergen1964.

[43] Cf. Stein Rokkan & Jean Meyriat, *International Guide to Electoral Statistics,* Vol. I: National Elections in Western Europe, Mouton, Paris 1969, "Denmark", pp. 59 f.

and this proportion was composed of about 73 % of all men above 30 years of age.

Turning to Sweden, we shall not go into details on the complicated and unstable social background of the Estate Assembly.[45] The right of the peasantry to elect their own Estate representatives had been officially established anew in 1720. But the *frälsebönder*–as well as the rural landless–were excluded from the Peasant Estate, and did not take part in the *Riksdag* elections.[46] Within this rough limit, each district regulated its own franchise, and there was no national minimum qualifications for the exercise of the franchise to the Burgher or the Peasant Estates.[47]

The two-chamber *Riksdag* of 1866 seems to have been comparatively less open to general peasant inclusion than the unicameral Norwegian *Storting* after 1814 and the Danish *Rigsdag* from 1849. The transition from Estate representation to general citizenship–after a constitutional struggle for several decades–implied no radical democratic extension of the franchise. The net increase of the rural electorate has been estimated at about 15 %, while about 1/10 of earlier voters of the Peasant Estate failed to satisfy the new suffrage qualifications.[48] The electoral colleges which selected the representatives to the First Chamber were subject to economic regulations, which gave the vote to all persons aged 21 and above who paid local taxes. Some economic qualifications were also operative as barriers restricting the franchise for elections to the Second Chamber.[49] This constitutional replacement of judicial Estate incorporation by linear economic barriers implied that about 5 % of the population had the right to vote in 1866. It has been estimated that in 1872 the proportion of enfranchised men of age was c. 22 %.[50]

44 The elections to the *Landsting* were indirect: half of the electors were those enfranchised for Lower House election, while the highest taxpayers constituted the other half, outside those 12 out of the 66 members who were appointed by the king. Cf. ibid., and Winding, "Danmark", in *Framveksten . . .*, Bergen 1964, op.cit., p. 81.

45 See Sten Carlsson 1949, op.cit., Ragnar Olsson 1926, op.cit., and Ragnar Olsson, *Riksdagsmannavalen till bondeståndet under den senare delen av frihetstiden (1740–1772)*, Lund 1948.

46 While the number of *frälsebönder* steadily declined, the proportion of rural landless increased substantially towards the middle of the 19th century, as we saw in chapters 5 and 6.

47 Cf. Douglas Verney, *Parliamentary Reform in Sweden 1866–1921*, Oxford at the Clerendon Press 1957, p. 92.

48 Cf. Wallin 1961, op.cit., p. 50.

49 Cf. Esbjörn Janson, "Sweden", in Rokkan & Meyriat 1969, op.cit., p. 283 and passim.

50 Cf. Kuhnle 1972, op.cit.–taken from "Bidrag til Sveriges officiella statistik", *Statistiska Centralbyråns Underdåniga Berättelse rörande Riksdagsmannavalen år 1872*, Val-statistik I (Stockholm 1873), p. 18.

Beyond the incorporation threshold, there were some systemic barriers implicit in the system of *representation.* This system defines some crucial conditions for the inclusion into the polity, and our survey will be confined to the major representative reorganizations after 1814, 1849 and 1866 for Norway, Denmark and Sweden respectively.[51] We will take a look at the eligibility requirements, the procedure and period of election, the question of electoral equality, and the chamber system of the assemblies, without any intention of presenting the rules in minute details.

The Norwegian constitution stated that the representatives had to be over 25 years of age, the same requirement as for franchise, and with the same property qualifications as for the electorate. The elections took place every 3 years, and followed an indirect procedure. The electors–about ten times as many as the final number of representatives–were chosen by the voters of the local constituencies in a proportionate number to the amount of enfranchised. The electors then came together to select the allocated number of representatives from the provinces and major cities–usually by open voting until the middle of the 1880s.[52] An increase in the number of registered citizens in a constituency should result in a proportionate increase of its number of electors and representatives, with an upper limit of 100 members of the *Storting*, and a lower limit of 75. But this was an unstable provision, and the socio-economic development itself broke the system: the upper limit had to be removed as early as 1842, and new electoral provisions were decided from 1859.[53]

The regulations of 1814 provided electoral equality within each constituency–by a one man, one vote principle–but there were gross inequalities across constituencies, to the systematic disadvantage of the rural areas. The cities were to have half as many representatives as the rural districts, which implied–according to our previous figures on urbanization (chapters 5 and 7)–that one vote in the cities weighed about as much as five votes in the countryside. At first this provision made it possible for the urban upper classes to counterbalance the overwhelming peasant majority by electoral overrepresentation. Later, when the greater cities grew rapidly and claimed representation proportionate to their number of voters, the clause came to be defended by the leaders of the mobilizing peasantry.[54] From the new electoral pro-

[51] These parliamentary systems are immediately comparable. The Swedish Estate Assembly (1718–1866) was, as we have seen, of a completely different quality, and the Danish Provincial Estates (1831–1849) were preliminary and consultative only.

[52] Cf. *Det Norske Storting gjennom 150 år,* vol. I, op.cit., p. 113 and p. 116. On this system, and later revisions, see Bjørn Kristvik and Stein Rokkan, "Valgordningen", in *Politiske valg i Norge,* stencil, Univ.forl., Oslo 1966.

[53] Cf. *Det Norske Storting,* vol. I, op.cit., pp. 73 ff.

[54] Cf. ibid., pp. 71 ff., and Rokkan in Lipset & Rokkan 1967, op.cit., pp. 386 ff.

visions of 1859 then, the 1:2 urban-rural ratio was strictly enforced, while each constituency was to send a fixed number of representatives to the *Storting* irrespective of changes in the number of registered voters.[55] The development is indicated by the following table:

Urban vs. rural representation to the Storting

Storting	Number of repr.	Urban repr.	Rural repr.
1815	87	c. 30 %	c. 70 %
1830	81	c. 37 %	c. 63 %
1842	100	33 %	67 %
1859	117	c. 43 %	c. 57 %
After 1859	111	$33^1/_3$ %	$66^2/_3$ %

Source: *Det Norske Storting gjennom 150 år*, vol. I, p. 74.

$33^1/_3$ %

It should be noted, finally, that the *Storting* was a unicameral legislature as far as the elections were concerned, so that no upper house modified the above picture as a 'conservative guarantee'.[56]

The Danish constitution of 1849 stated that the number of representatives to the *Rigsdag* should be fairly proportionate to the total population, at a rate of about one for every 14 000 inhabitants. The eligibility requirement was 25 years of age for the lower house, the *Folketing* (while no men under 30 were given the vote, as we saw above), and 40 years for the *Landsting*. There were no property qualifications for *Folketing* representatives, while the *Landsting* members had to have a minimum amount of direct tax or assessed income. The constitutional revision of 1866 still restricted upper house membership to the higher income groups, and the king also got the right to appoint 12 of the 66 representatives.[57] The election period to the *Folketing* was 3 years, with direct elections by simple majority in single-member constituencies.[58] The *Landsting* employed an election period of 8 years, with half of the body up for election every 4 years. In 1849–1866 the elections were indirect, with a primary choice of electors to

[55] The same provision–the 1:2 ratio–which had served to ensure urban interests in 1814, had become a defence of rural representation in an increasingly urban society, and came to be called the 'peasant clause'–*bondeparagrafen*. See, inter alia, Koht, "Bondeparagrafen i historisk lys", in *Nye innhogg og utsyn*, Aschehoug, Oslo 1964, pp. 154 ff.

[56] See *Det Norske Storting*, I, op.cit., pp. 171 ff. on the organization of the *Storting* and its division into a *Lagting* and an *Odelsting*.

[57] Cf. *Den Danske Rigsdag*, Vol. I and II, passim.

[58] There was usually open voting–by show of hands–until 1901.

the regional electoral colleges. From 1866 the procedure was rather complicated. Twelve members, who should have been representatives on previous assemblies, were appointed for life by the king, and one was elected by the Faroese *Lagting*. The remaining 53 were chosen indirectly by electors. Half of the electors were elected by a plurality of those enfranchised for *Folketing* elections, while the other half was selected by the highest taxpayers.[59] Proportional representation (PR) was employed for elections to the upper house.

As concerns the degree of electoral equality, the principal built-in constitutional bias aimed at securing landowner control through the *Landsting*, particularly from the reversal of 1866. Those conservative barriers against popular influence which the urban-rural representation ratio was originally to secure in Norway, were provided for by economic restrictions and the modelling of the chamber system in Denmark. But the electoral inequality intrinsic to the character and position of the *Landsting* were not in principle based on urban representation. Barriers against rural representation would in fact have affected the great landowners, and–owing to an agrarian structure different from the Norwegian one–another type of check on the representative system was established.[60]

The Estate system in Sweden had become increasingly outmoded from the early 19th century. The Estates of the *burghers* and the clergy lost some of their original basis and identity, while farmers owning tax-exempted land (*frälsehemman*) were admitted to the Peasant Estate from the 1840s.[61] The elections to the Estate Assembly had usually been indirect, in single-member constituencies. The mimimum age for members in any Estate was 21 years, and no one eligible for election to another Estate was allowed to represent the Peasantry in the *Rigsdag*. The Parliament Act of 1866 introduced new principles for representation. Eligibility to the upper house required the candidate to be at least 35 years of age, with an assessed property or an assessed income of a minimum value.[62] Representatives to the lower house had

[59] In the rural districts the highest taxpayers–mostly the greater landowners–were automatically the electors, while an indirect procedure was employed for Copenhagen and the provincial towns. Cf. Rokkan & Meyriat 1969, op.cit., on "Denmark", p. 60, and Erik Rasmussen, *Komparativ Politik*, vol. 2, Gyldendal, Cph. 1969, pp. 130 f.

[60] Which is not to say, of course, that the agrarian structure was a sufficient cause of the more minute differences in constitutional arrangements. On the regulation of constituencies etc. in the Danish constitution, see the constitutional texts, as appendices to *Den Danske Rigsdag*, vol. I and II, op.cit.

[61] The Burgher Estate changed its composition with the enfranchisement of property-holders in 1858–which meant the abandonment of profession as an exclusive qualification while the learned professions had in 1823 already obtained admittance to the Estate of the Clergy. Cf. Verney 1957, op.cit., pp. 31 ff. On the obliteration of the Swedish Estate society generally, see Sten Carlsson 1949, op.cit.

[62] The property was to be worth at least 80 000 *riksdaler*, and the income 4 0000 *rdr*.

to be at least 25 years of age, while the property qualifications were similar to those governing the electorate.[63] Elections to the upper house were made indirectly–by the provincial and city councils–and the period for each representative was 9 years, with a fraction elected annually. The lower house representatives were elected for 3 years, either directly or indirectly, and mostly by absolute majority in single-member constituencies.[64]

As for the degree of electoral equality, the principle of equal suffrage (one man, one vote) was applied to the lower house, while a system of plural voting governed the upper house elections. Since the representatives were elected by the provincial and city councils, this system followed the income-graded franchise scale which applied to the municipal elections.[64b] The graded scale reinforced the social inequality implicit in the rather severe demands of income or fortune for eligibility qualification.

The principal constitutional check on popular influence in Sweden from 1866 was thus the two-chamber system with an aristocratic upper house. Both chambers had the same tasks and authority, and since they were able to paralyse each other, the system implied a built-in counterforce to the peasant majority in the lower house.

This brief comparative survey of the development of political systemic contexts is condensed into a summary chart at the end of the present section.[65]

Relations of authority

The sequence of thresholds outlined above is also an expression of transitional relations of authority between social groups (cf. chapter 5). The Norwegian peasantry attained political citizenship by the May constitution of 1814, and the Danish by the June constitution of 1849. But this right remained a privilege connected to the possession of land. Like the Swedish Estate system, suffrage and eligibility were confined to the freeholders and the independent leaseholders, and excluded the rural landless together with the urban proletariate. Reminiscences thus remained of an Estate society–ideal-typically based on absolute principles of classification, discontinuous scales of rank, and judicial-

63 Leased property should be assessed at minimum 6000 *rdr.*, or owned property at 1000 *rdr.*, or annual income at least 800 *rdr.* Cf. Verney 1957, op.cit., pp. 248 f.

64 Cf. Herlitz 1964, op.cit., pp. 237 ff.

64b The number of votes was unlimited in the country, while there was a maximum of 1/20 of the total amount of votes in the cities from 1866, and 1/50–or 100 votes–from 1869. Cf. Janson in Rokkan & Meyriat 1969, op.cit., p. 283.

65 The chart is a composite of categories from Rokkan 1970, op.cit., pp. 84–85, Lafferty 1971, op.cit., fig. 4.1, pp. 166 ff., and my own account.

political assignment of social status.[66] The Norwegian peasantry were even able to retain the allodial privilege (*odelsrett*), as a protection against more complete mercantilization of land. This privilege represented the major institutional defence of landholding peasanthood in relation to the dual capitalist expansion–through the rural landless as well as the commercial upper classes. There was nevertheless, as may be concluded from chapters 6–7, an increasing generalization of monetary criteria as a common societal denominator. Politically, this is expressed by economic qualifications for enfranchisement, which –in one form or another–dominated all the Scandinavian electoral systems during the last quarter of the 19th century. At the initial moment of political incorporation, though, the Scandinavian peasantry were still charged with specific duties as a societal Estate. They had a general obligation of conveying travellers of superior rank, and–above all–they were exclusively liable for military service. These onerous hallmarks of institutionalized subordination became rather dominant issues in peasant mobilization towards the middle of the 19th century, as we shall see later.

Summary Chart of Political System Development in the Scandinavian countries from the late 18th century to c. 1870. Emphasis on Peasantry

Features of national integration and political thresholds	Norway	Denmark	Sweden
I. *The Regime*			
1. Type of state formation	Dynastic Monarchy	Dynastic Monarchy	Dynastic Monarchy
2. Inherited style of rule	Monarchic absolutist	Monarchic absolutist	Continuous representative
3. Initial organ of representation	Storting 1814	Provincial Estates 1831–1849	Estates Assembly
4. Major Representative reorganization		Two-chamber Rigsdag 1849	Two-chamber Riksdag 1866

[66] Cf. the lucid article by Ottar Dahl, "Noen teoretiske problemer i sosialhistorien", *Historisk Tidsskrift*, vol. 37 (1955), pp. 185–203, where this type of social formation is contrasted with a 'class society' characterized by 'gradual' principles of classification, 'continuous' scales of rank, and status more explicitly expressed by economic power. The article is reprinted in *Hundre års historisk forskning*. Univ.forl., Oslo 1970.

Features of national integration and political thresholds	Norway	Denmark	Sweden
5. Parliamentery Control based on Lower House majority	1884	1901	1917
II. *Legitimation threshold*			
1. Local landlord jurisdiction	None	Until the Peasant Liberation of late 18th century.	Partially on noble lands before 1680 – mainly confined to the southern counties
2. Tradition of tolerance of peasant opposition before major representative reform	Petitions directly to the King	(Local jurisdiction and patronage)	Petitions through the Estate Assembly
3. Last major anti-peasant suppression	Peasant march against the Storting stopped by armed force in 1818. Leaders of rural proletarian movement imprisoned in early 1850s.	Freedom of (rural) assembly suspended in 1845	Public suppression of movements among the dependent *frälsebönder* in 18th 19th centuries (cf. the text)
III. *Incorporation threshold*			
1. First national election with peasant candidates	1814	(1834–1849) 1849	(1718–1866) 1866
2. Per cent of population enfranchised	1814. c. 10 % 1870: 7.4 %	1834: c. 3 % 1849: c. 15 % 1866 (Reversal Upper House.)	1866: c. 5 %
3. Per cent of men above minimum age enfranchised ..	1814: c. 45 % 1845: c. 35 % 1875: c. 34 %	1849: c. 73 %	1872: c. 22 %

4. Institutionalized electoral inequalities	Mixture of estate, property and competence reg.	Income or property qual. for Upper H. from 1866	Econ. qual. for both chambers. Plural voting for Upper House
IV. *Representation threshold*			
1. Upper House (Denmark and Sweden)			
a) Eligibility requirements			
i. Age		1849: 40	1866: 35
ii. Property qualifications		1849: Assessed income of 24 000 *Kr.* or dir.tax of 400 kr.	1866: assessed estate of 80 000 Rdr. or ass. inc. of 4000 rdr.
b) Election period		1849: 8 years	1866: 9 years
c) Procedure of election		1849: Indirect (regional electoral colleges) (PR) 1866: 3/4 indir. 1/4 app. by King	1866: Indirect (provincial and city councils)
2. Lower House			
a) Eligibility requirements			
i. Age	1814: 25	1849: 25	1866: 25
ii. Property qualifications	1814: Same as electorate	1849: None	1866: Same as electorate
b) Election period	1814: 3 years	1849: 3 years	1866: 3 years
c) Procedure of election	1814: Indirect; absolute majority	1849: Direct: single-member districts; simple majority	1866: Dir. and indir.; absolute majority
d) Major barrier against equal vote representation	1 : 2 urban-rural ratio	Aristocratic counter-balance with strict econ. qual. and royal appointment of members	Aristocratic counter-balance by Upper House, with strict econ. qual. and graded plural vote

Political participation

Even if the Scandinavian peasantry passed the threshold of formal incorporation into the national polity, this is not to say that they exerted their potential influence, or took part in the macro-political contest to the same extent as their counterparts from the higher non-peasant classes. On the contrary, this was not even expected by their urban and educated betters. Substantial parts of the peasantry showed a very low standard of education during the first half of the 19th century, and the first primary school laws, which required regulated country-schools and some minimum qualifications for teachers, became not much more than pipe dreams in many areas until well into the second half of the century.[67] Communications were also generally bad, and the peasantry had no political training. Three aspects of the question of the overall political participation of the peasantry will be outlined below: the average figures of participation in the national elections after the introduction of constitutional regimes; the major pattern of peasant voting; and–in the subsequent section–the rough pattern of participation in local government.

As for Norway, the development of the general level of participation at the polls is shown by the following table.[68]

This table includes some figures for the major waves of rural electoral mobilization–first at the election in 1833, then again from the 1870s to the culmination with the constitutional struggles of the early 1880s. Turnout remained on the average fairly stable during the 1840s, while the level declined in the 1850s.[69] The lowest poll was in the election at the middle of the 1860s, when one of three men was qualified for the vote, one of five was registered, and only one of ten actually took part in the election.[70] The major demonstration in the table below, however (and the reason why we find it unnecessary to complete it with all elections during the period), is the generally low participation at the

67 A comparative account of education during the period investigated here is presented in B. J. Hovde, *The Scandinavian Countries 1720–1865,* 2 vols., Chapman & Grimes, Boston 1943, ch. XV, pp. 589 ff.

68 The first official electoral statistics of Norway are presented in *Valgmandsvalgene og Stortingsvalgene 1815–1885,* NOS III, nr. 219. Some figures here, particularly for the elections until 1826, are inaccurate estimates. For further comments on turnout and electoral practice 1814–1869, cf. *Det Norske Storting,* I, op.cit., pp. 89 ff. and pp. 110 ff.

It should be noted that "cities" in the table comprise urban districts under royal charter, the *kjøpsteder,* while "rural" includes both market towns (*ladesteder*) and rural communes.

69 Cf. *Historical Statistics 1968,* Oslo, op.cit., p. 631.

70 These estimates are based on information presented in the preceding section. See also Jostein Nerbøvik, *Norsk Historie 1870–1905,* Oslo 1973, pp. 53 f.

Percentage poll in Norway. Elections to the Storting

Year	Votes cast in per cent of qualified *cities*	*rural* (no data until 1859)	*total*	Votes cast in per cent of registered *cities*	*rural* (no data until 1829)	*total*
1829				64.8	43.0	45.3
1832				59.3	46.8	48.0
1835				63.6	51.0	52.2
1838				64.8	48.6	50.0
1859	40.2	29.5	31.0	60.2	45.6	47.6
1862	33.7	28.1	28.9	52.0	43.4	44.6
1865	29.1	26.4	26.8	46.2	41.1	41.8
1868	29.9	30.5	30.4	47.8	47.6	47.6
1870	23.8	29.5	28.7	40.3	45.8	45.1
1879	40.0	28.4	30.4	65.5	45.2	48.7
1882	61.5	46.9	49.5	82.6	70.1	72.5

Sources: Compiled from *Historical Statistics*, Oslo 1968 table 360, p. 631, *Det Norske Storting gjennom 150 år* I, p. 113, and Rokkan in Lipset & Rokkan 1967 p. 378.

national elections, and the even substantially lower turnout in the rural areas compared to the cities.[71] This fact cannot as such be taken as a valid indicator of general political apathy among the rural population, but it is an expression of the relatively limited grass-roots mobilization behind peasant politics in the parliamentary arena of power.

The degree of actual agrarian mobilization, relative to the substantial numerical dominance of the peasant electorate, is further indicated by patterns of voting and representation. The following table shows that the governmental officials dominated the *Storting* in the decades after 1814, and constituted the largest group until the "peasant Storting" (*bondestortinget*) of 1833. But their position on the whole declined throughout the period, parallel to the increasing peasant support of representatives from their own class, while the percentage of *burghers* remained fairly stable from 1814 to the 1870s, with a peak in the 1820s and one in 1851.

[71] The percentage poll in the countryside did not approach the urban figures until the 1890s, and the 'nationalized' political contest does not seem to have become relevant to some peripheral areas until several decades into the 20th century–cf. Stein Rokkan and Henry Valen, "The Mobilization of the Periphery", in Rokkan 1970, op.cit.

Social composition of the Storting *1815–1874. Percentages*

Year	Public officials	Burghers	Peasants, sheriffs, parish clerks	Attorneys and others
1814	46	12	36	6
1818	54	12	23	11
1821	42	23	32	3
1824	46	20	31	3
1827	51	18	25	6
1830	48	17	25	10
1833	34	15	47	4
1836	38	12	45	5
1839	46	13	35	6
1842	41	9	43	7
1845	43	12	41	4
1848	43	11	37	9
1851	26	18	41	15
1854	38	11	41	10
1857	35	13	38	14
1859	28	15	40	17
1862	33	12	43	12
1865	32	11	47	10
1868	27	13	50	10
1871	20	11	59	10
1874	26	10	48	16
1814–27	47	17	30	6
1830–42	41	13	39	7
1845–57	37	13	39	11
1859–74	28	12	48	12
1814–74	37	13	40	10

Source: J. N. Mohn, "Storthingene 1815–1874, belyste ved Statistik", *Norsk Retstidende*, 1874, pp. 763 f.

The *burghers* represented mainly the urban constituencies, while a substantial amount of the governmental officials were elected from the rural districts, as shown by the following table:

Representatives from the rural districts. Percentages

Year	Public officials	Peasants, sheriffs, parish clerks	Attorneys and others
1814–27	47	45	8
1830–42	31	59	10
1845–57	29	63	8
1859–74	17	74	9
1814–74	29	62	9

Source: Mohn 1874, op.cit., p. 771.

The peasantry (together with the sheriffs and parish clerks, who usually were peasants themselves), gradually increased their share of the rural representatives from–on the average–less than half in the first period to about 3/4 in the last one. The peasants proper (sheriffs etc. excluded) made up for about 1/3 of the representatives of the whole *Storting* during their peak periods in 1830–42 and 1859–74, and 1/4 in 1815–27 and 1845–57.[72] The increasing peasant mobilization at the nominations in the countryside might thus be an expression of a 'class' consciousness which–as we shall consider in chapter 9–implied a growing tension between the urban centres and the rural periphery.

In Denmark, the electoral participation relating to the first provincial assemblies of Estates seems to have been remarkably high–c. 83 % on the islands and c. 87 % on Jutland. But this must be seen against the background of more severe landholding qualifications for suffrage than in Norway, and the percentage also fell somewhat at later elections.[73] The June Constitution (1849) implied a radical extension of the franchise, and the turnout at the elections was very low–on the average 25–30 % for the period 1849 to 1866.[74] The percentage poll relative to the electorate then rose steadily after 1866–being a good 30 % in

[72] J. N. Mohn 1874, op.cit., pp. 767 f. Cf. also the comments on the composition of the *Storting* in *Det Norske Storting,* I, op.cit., pp. 146 ff., and in Rokkan, Lipset & Rokkan 1967, op.cit., p. 370.

[73] Cf. Kjeld Winding, *Den Danske Bondebevegelse,* Det Danske Selskab, Copenhagen 1943, pp. 11–12.

[74] It seems to be a general phenomenon that turnout declines immediately after new groups are given the vote. Politically marginal and non-mobilized groups were involved in the nearly universal manhood suffrage; and the rural population often had a long way to go to the voting place. For further details on voting turnout during this period, see generally *Den Danske Rigsdag 1849–1949,* vol. III, p. 72 and passim; and Erik Høgh, *Vælgeradfærd i Danmark 1849–1901,* Copenhagen 1972.

1866 and 1869, a good 40 % in 1872, and a good 50 % in 1875. The intense constitutional struggle between the Right and the Left was probably an important mobilizing factor, since basic interest-accumulating prerequisites like general education and newspapers were also more widespread now.[75]

Around 1860, the highest voting percentage was to be found in Zealand. As we shall see in chapter 9, the organization of the "Friends of the Peasants" (*Bondevennerne*) had its stronghold here, and the numerical peasant predominance became early visible in the parliamentary representation.[76] On Jutland, on the contrary, the peasants more often chose a public official or a big landowner to represent them. The more thorough transition towards freehold on Jutland probably meant a less antagonistic relationship to the landed lords, and the Jutland peasantry were thus somewhat alienated from the program and alliance of the peasant movement on the islands. Jutland was, accordingly, mobilized for the liberal peasant party, *Venstre,* later than Zealand.[77]

Since our period of investigation ends at about 1870, we shall not go into much detail on the development of electoral participation after the representative reorganization in Sweden (1866). The average turnout at the polls is given in the following table, where the direct and the indirect elections are grouped together.

Percentage poll in Sweden 1866–1881
Second chamber elections

Year	Cities – all elections	Rural – all elections	Total
1866	36–37	15–16	c. 18
1869	c. 45		
1872	41.1	16.1	19.1
1875	36.7	16.7	19.5
1878	39.5	16.7	20.3
1881	45.2	19.5	23.7

Source: Wallin 1961, p. 386.

These figures indicate that the turnout was particularly low in the rural districts, and that the participation remained at a stable low level from the first election and throughout the 1870s. This might generally

[75] The poll decreased slightly during the latter part of the 1870s, to increase steadily again in the following decade – cf. vol. III ibid., p. 89 for further details.

[76] Cf. Kjeld Winding, "Valgkredserne gennem tiderne", in ibid., p. 12.

[77] Ibid., pp. 12 f. See the outline in our subsequent chapter on the peasant movement.

be explained by the usual factors: few and scattered voting places, bad communications, a low standard of education. It should be noted, however, that the voting turnout in the countryside was about double at the direct elections compared to the indirect ones – but even then the participation at the polls seldom surpassed a quarter of those qualified for the vote.[78] Despite the low level of participation, the proportion of peasant representatives was already fairly high from the first Second Chamber election, and it rose during the first decades afterwards. In 1866 the social composition of the Second Chamber was as follows:[79]

Agriculturalists	98
(Peasants of these	75)
Officialdom and liberal occupations	61
Industrialists	20
Merchants and bankers	11
Total	190 representatives

The peasantry thus accounted for nearly 40 % of the representatives, and before long they controlled a majority in the Second Chamber.

Generally, the Scandinavian peasantry at the grass-roots level do not seem to have been overwhelmingly interested in the national elections before the last quarter of the 19th century. The figures are not immediately comparable, since neither the constitutional frameworks nor the suffrage and eligibility qualifications were identical. It has therefore been unnecessary, and in fact quite impossible, to present an equally balanced account for each of the three polities. In Norway there was, however, a notable process of rural mobilization from the 1830s, while the situation in Denmark and Sweden was more obscure, since we are not able to judge what the transition from (different) Estate constitutions to two-chamber assemblies actually meant in this respect. We will approach the problem through other types of evidence below.

Local government and peasant participation

The preceding outline immediately provokes some crucial questions. Even if the peasant participation in national politics generally was low, there existed other channels of participation at the local level. How was the peasantry situated in relation to various types of local government? Did there exist official political structures which were more in

[78] Cf. Wallin 1961, op.cit., pp. 386 ff.

[79] Cf. Edvard Thermaenius (ed.), *Svensk Bondepolitik*, Gebers förlag, Stockholm 1931,

accordance with the structure of the peasant society itself–corresponding to its–relatively–communalistic orientation? More explicitly:
Was local self-government opened up as a field of peasant activity which diverted them from entering national politics with greater force? We will consider these questions in a short comparative survey, without aiming at definite answers.

In the twin kingdom of Denmark–Norway, decisions on local affairs had been taken by the central or provincial officialdom, although rudimentary institutions like peasant village meetings or parish councils –as reminiscences of the old Nordic traditions of folk-assemblies– sometimes might have been consulted.[80] This autocratic system continued in the local communities of Norway for more than two decades after 1814, while the peasants had a right of representation at the national level. But a reorganization of local government was officially prepared, in order to simplify the communal administration and clarify the division of competence between the local communes and the administrative units at the level of the county (*amt*) and diocese (*stift*).[81] The peasantry, and their representatives in the *Storting*, at first showed a remarkable indifference towards reforms of local government, but they seem to have discovered the possibilities of communal self-government at the *Storting* in 1833: this was not a theoretical issue, but eminently a practical tool in the struggle against the bureaucracy. Matters like the poor relief system, and questions of schools and public health, implied increasing claims by the government. The peasant motive was to keep the state apparatus in check, to transfer the burden of taxation to the monetary urban economy, and to keep public expenses in the peasant communities down. The provincial officialdom could be directly controlled by means of local self-government. The peasantry on the *Storting* of 1833 then advocated a more radical reform than anybody else–implying a system of 'communal sovereignty', with the peasant communities as rather autonomous units outside supra-local administrative control. The legislation on local government of 1837 grew out of the conflict between the official conception of self-government combined with administrative authority, on the one hand, and the peasant idea of decentralized and autonomous communes on the other.[82] The basic representative organ within each local district was

80 See Hovde 1943, op.cit., pp. 567 ff. Hovde quite obviously overestimates these features however.

81 Cf. Steen, *Lokalt selvstyre i Norges bygder,* Cappelen, Oslo 1968, pp. 13 ff.

82 Ibid., pp. 9 ff.; cf. also *Det Norske Storting gjennom 150 år,* vol I, pp. 501 ff., and Steen, *Det frie Norge,* vol. V, pp. 288 ff. The radical bill that was adopted by the peasant representatives in 1833 was opposed by the ministry and vetoed by the King. A royal commission appointed in 1834 then prepared a bill which was supported by the peasants and the liberals, and which received the royal signature in 1837.

to be a council, which was elected for 4 years by suffrage regulations like those which applied to the *Storting* elections, with half of the body subject to election every second year. The council chose an executive committee and a chairman, and the local chairmen together constituted the executive committee of the county.[83] The council was authorized to perform economic decision-making and local taxing and budgeting with a limited right of approval for the county governor or the king.[84] Nevertheless, the turnout–in percentage of the registered electorate, was quite low at the first local elections. It was, on the average, below the participation at the polls in the *Storting* elections, and in 1837 it varied geographically from 15 to 60 % of the registered voters. It was even lower during the 1840s and 1850s–from 10 to 30 %–and did not reach the turnout level of the first election until the 1860s.[85] It has been established, though, that the proportion of actually *registered* voters, from the ranks of those qualified for the vote, increased substantially in many rural districts after the parliamentary election in 1836. This fact might indicate an anticipating interest in the potential of participation at the local election, even if it was not sufficient for actual voting on the election day.[86] It thus seems that the rural grass-roots electorate did not acknowledge the significance of the representative local institutions to an extent that measured up to the interest shown by the peasant leaders in the *Storting*.

The peasant failure to fully exploit politically their numerical dominance, in contrast to the local officialdom, is also shown by the elections of council chairmen in the rural municipalities.[87] According to *Fladby*'s estimate, the percentage of peasant chairmen in the rural communes runs as follows:[88]

1838–42.5 %; 1844–58 %; 1850–68 %; 1856–65 %; 1862–c.73 %.

Nearly a third of the local chairmen were parish priests, and a tenth were officers, in 1838–a fact which clearly indicates the position of patriarchal officialdom in the peasant communities. There were local

[83] Cf. Steen, *Amt og stat 1837–1860,* Cappelen, Oslo 1973.

[84] It should be noted that the new system definitely needed some running-in period before it was carried out all over the country– cf. Steen 1968, op.cit., passim.

[85] Steen 1968, op.cit., p. 23. It should be noted, however, that all elections after the first one in 1837 were only supplementary, for a half of the council. This fact had probably in itself some de-mobilizing effects.

[86] Steen 1968, op.cit., pp. 21 ff.

[87] See Rolf Fladby, "Bønder og embetsmenn i lokalstyringen etter 1837", *Historisk Tidsskrift,* vol. 46 (1967), pp. 21–53.

[88] Ibid., pp. 26 ff. Fladby's estimate is partly based on a list compiled by Arne Bergsgård, in "Formannskapslovene, opphav og utvikling", *Minneskrift til formannskapslovenes hundreårsjubileum,* 1937. Steen, 1973, op.cit., pp. 48 f. points out that the list of chairmen is still incomplete even after being supplemented, particularly for the county of Finnmark in the far north. But this definitely cannot change the overall picture.

variations, with most peasant chairmen in the more peasant-dominated areas, and in the counties of the south and south-west. The overall peasant proportion of local chairmanship also increased to nearly 3/4 in the 1860s. But still the provincial officialdom, and particularly its clerical branch, was disproportionately dominant in local public affairs after the radical reform of 1837. This fact should not, however, obscure the undisputable and steady increase in peasant mobilization in the decades after the 1830s, where local politics often served as an apprenticeship for national politics, and not only as a substitute for it.

The reorganization of autocratic local government in Denmark was also prepared during the 1830s. The issue was–according to the intention of the monarchic government–to be handled by the provisional system of consultative Estates, and the matter raised conflict in the assemblies.[89] The principal regulations for the rural communes were decreed in August 1841. Parish councils with particular care for school matters, road maintenance, poor relief and public health were established in the rural districts, and the pastors remained members *ex officio,* together with the greatest landowners. The right to vote was given to those who possessed a minimum amount of land.[90] Thus the landowners definitely dominated, although the peasantry were also included, and gained some competence in local politics. The system was revised by a new law in 1855, which still retained the principle of class voting, whereby the highest taxpayers were to elect a majority of the council members, although *ex officio* membership was abolished. Peasant autonomy–actual or potential–within the local setting, on the relatively democratic Norwegian model, was not feasible on Danish lands, and the system of local government here expresses dissimilar structural conditions.

In Sweden, the development of types of self-government for the villages from the late 18th century had been restrained by the enclosure movement, whereby the old village communities were broken up (cf. chapters 5–6). The relatively high autonomy of the communes (*herreder*) formally endured, but was gradually transferred to the parishes –particularly after the law of 1842 which required a primary school with a trained teacher in each parish. The constitutional framework of the parish had become regulated by decrees of 1817 and 1843. The first one reserved the vote at the parish meetings for landowners, and retained a graded vote in proportion to registered landed possession (the *mantal*), while the second one established an executive committee

89 Cf. Hans Jensen, vol. II, 1934, op.cit., pp. 474–485. Cf. also Hovde 1943, op.cit., pp. 569 ff.

90 *Ex officio* membership of the council required ownership of 32 *tdr. hartkorn* while 1 *tdr.* was enough for franchise. Cf. Aksel Milthers in K. Hansen, vol. V, op.cit., p. 257.

for secular matters parallel to the ecclesiastical council.[91] But communal self-government, as a training-ground for citizenship, was increasingly a liberal issue during the 19th century, and the problem of reform in local government was discussed at nearly every *Riksdag* session after 1823. The new framework was defined by the regulations of 1862. The parish (*socken*) was transformed into a rural commune (*landskommun*), and the vote was weighted on a graded scale according to income or property instead of the registered *mantal*.[92] But this principle in fact implied that dominance by the landholding peasantry became less inevitable than before, since the vote for the representative assembly was no longer graded with the proper possession of *land*. The Swedish peasantry had never, though, experienced the formally more egalitarian local democracy of the Norwegian type. Compared to Norway, the Swedish as well as the Danish model of local government was based on institutionalized aristocratic predominance. But the Swedish reorganization of 1862 is significant for its abandonment of the outmoded Estate principle, whereby it anticipated the two-chamber *Riksdag* of 1866.[93]

Recapitulation

The constitutional reorganization in Norway (1814) implied general peasant citizenship within the polity much earlier than was the case in the other Scandinavian countries. Dominant sections of the Swedish peasantry were politically incorporated as a societal Estate, through a system of co-optation without much influence. In Norway there were no landed intermediaries between the peasantry and the bureaucratic elite, so if the new regime required legitimation from predominant sections of an agrarian society, the freeholding peasantry could hardly be ignored. This was equally the case in local affairs, particularly since peasant localism increasingly gave priority to the issue of self-government. The demand might have been even easier to comply with since local self-government subsequently diverted attention from the national arena, as indicated by a temporary decline of peasant mobilization to the *Storting*. Political mass participation was not very high, however, but it was nevertheless increasing all over Scandinavia, parallel to more exclusive voting for peasant candidates.

In Denmark there was a fairly radical political inclusion of the landholding peasantry from 1849, but this reform was basically instrumental to the liberal bourgeoisie (cf. chapter 9 below). The landowning

91 Cf. Herlitz, 1964, op.cit., p. 212 d.
92 Herlitz 1964, op.cit., pp. 214 f. Cf. also Ingers, vol. III, op.cit., 302 ff.
93 Cf. Sten Carlsson 1949, op.cit., pp. 338 ff.

aristocracy retained their dominance in the local setting, and strengthened their national position by the constitutional reversal of 1866. This occurred simultaneously with the abolition of the Swedish Assembly of Estates. Here the principles of judicial and discontinuous assignment of political rank were replaced by economic qualifications which equally secured aristocratic predominance, and left out the smallholders and the landless altogether. It was, however, a general Scandinavian phenomenon that monetary criteria increasingly replaced assignment of political status based on landholding–in accordance with the gradual penetration of 19th-century agrarian society by commodity relations and monetary principles.

In the subsequent chapter we will consider how the non-uniform features of essential peasant politics–given the available options within the political systems–hinge on the interrelated juxtaposition of agrarian structure and socio-economic change.

CHAPTER 9

The peasant movement

We have seen that the Scandinavian peasantry were gradually included in the national polities during the 19th century–in terms of formal status as well as political behaviour. But these are not the only aspects of the development of peasant politics. The character of peasant demands–towards the state power and towards superior social classes–also underwent a fundamental transformation during the period in question. We will have to trace two types of problems–still in a conventionalized and comparative manner:

Firstly, the intrinsic character of peasant politics within each national setting–its aims and channels of activity, and its 'ideological' legitimations and degree of constitutionalism. Secondly, the pattern of political cleavages *within* the peasantry, concentrated on the questions about the conditions under which such splits occurred, and what types of cleavage dimensions they revealed. The implications for the overall national alliance patterns will be explicitly discussed in the final section of this chapter. For the sake of descriptive lucidity, we will first give a coherent survey for each country successively. The account will be analytically related to that structural transformation which was outlined in Part Two.

Norway

The Norwegian peasantry under absolutist rule–prior to 1814–had, as we saw in chapter 8, a long tradition of petitions and expeditions of complaint direct to the supreme power–the King in Copenhagen.[1] To the peasant mind, the king often seems to have been perceived as a guarantor of legal rights against injustice from local power groups–including the provincial officialdom. He was often used as a lever in an elementary social struggle, and he probably demonstrated some compliance towards the peasantry, to counterbalance the economic-

[1] See Koht 1926.

political elite in distant Norway.[2] It was also more advantageous to the Crown to defend the peasantry as directly taxable subjects, instead of letting them pay rents to a class of landed lords. As suggested in chapter 8, the tradition of direct complaint served to stabilize the systemic balance, since the peasant voice was both a harmless outlet of frustration, and a confirmation of monarchic power as a factor of national integration. Local letters of complaint were thus commonplace under absolutist rule, whether they expressed peasant displeasure with heavy taxation, with an obstinate bailiff, or with a cheating merchant.[3] During the second half of the 18th century, however, there also occurred more dramatic peasant uprisings, which point beyond the traditionally restricted localism. These episodes–if so they were–should briefly be considered, because they expose latent antagonisms, and put the basic socio-political cleavages into focus.

The imposition of an extraordinary tax in the twin Kingdom in the 1760s–to supply the public purse after prolonged warfare–provoked notable peasant unrest in some western districts of Norway.[4] The peasantry of several poor seaside communities of western Norway complained of the extra tax in the early 1760s, and they usually combined these petitions with complaints of economic conditions and of undue charges and fees to the local officials.[5] Another strategy was passive resistance by refusal to pay the tax, and the bailiffs were supported by armed forces, at arms length, in their collection of duties in some areas at the middle of the decade. Some few hundred poor peasants gathered in Bergen in the spring of 1765. They confronted the provincial officials to get their money back, but had to leave the town

[2] One should be careful not to stretch this argument too far. It should also be noted that it was stated in a decree as early as the middle of the 17th century that complaints should be considered by the county governors, although the peasantry continually defied this prohibition of direct appeals to the king, so it was not actually enforced. Cf. Koht 1926, ch. 12, pp. 152 ff.

[3] Ibid.

[4] On the 'Strile war' (*Strilekrigen*)–named after a rural area north of Bergen–and parallel peasant reactions, see generally Steen, *NFLH*, vol. VI,pp. 390 ff., and Koht 1926, ch. 21. A new study of the implications of the extraordinary tax has recently been made by the Norwegian historian Gudmund Sandvik,–see *Det gamle veldet,* Gyldendal, Oslo 1975.

The tax levy required all persons above 12 years of age to pay 8 shillings a month or 1 *riksdaler* a year–which probably meant that the tax burden of the peasantry was doubled–cf. Steen, op.cit., p. 391.

[5] Cf. Egil Harald Grude, *Skatt, bønder og embetsmenn. Ekstraskatten i Stavanger amt 1762–65*, unpublished thesis in history, University of Oslo 1972, pp. 76 ff. The leaseholding peasantry of the western fishing communities might have been particularly depressed, since they were not able to supply themselves with sufficient bread corn, while corn prices rose, as did the value of land and thereby the leasehold duties–according to Koht 1926, pp. 291 ff.

when opposed by urban mercenaries and armed forces.[6] Similar peasant gatherings also occurred in certain areas further south–in the county of Stavanger–some months later, accompanied by tax refusals and letters of complaint to the king.[7] There were no extensive public sanctions against the oppositional peasantry, although the collection of the extra tax continued to be far from effective, so the tax was removed a few years later.[8]

Three characteristics of this event are particularly relevant to our analysis:

- the peasant unrest was–although supra-local–regionally restricted.
- it was a rather epidemic *ad hoc* movement, without traces of concerted action over a longer period of time.
- it was directed against the tax-collecting state in general, and the local officials in particular.

This cleavage–the peripheral peasant interests against the central polity and its provincial representatives–was not the only one in the old society. The most dramatic peasant uprising in the absolutist age was primarily triggered off by mercantilistic regulations and commercial privileges to the benefit of the urban merchants. From about the middle of the 1780s, a movement of peasant complaints spread in the southern counties of Norway.[9] The southern peasantry were increasingly engaged in subsidiary occupations like seafaring, lumber trade, and handicraft, and took an interest in more liberal market circulation of goods, against the commercial monopoly of the urban *burghers*.[10] The Danish monopoly on grain export to southern Norway from 1735 (cf. chapters 5 and 7) deprived the peasantry of free purchase of corn, and the cities had exclusive privileges on lumber export by foreign ships. This economic demobilization of the peasantry, together with rising prices in the last quarter of the century, implied increasing peasant indebtedness to the benefit of the commercial class, while the collection of claims was to be enforced by the public officials.[11] New tax regulations from 1781 imposed a ground tax in Norway, which implied further burdens on indebted freeholders, since the commercial capital

[6] Cf. Koht 1926, pp. 294 ff.

[7] Cf. Steen, *NFLH*, VI, pp. 393 ff., Grude 1972, op.cit., pp. 1 ff. and passim., Koht 1926, p. 299.

[8] Cf. Steen, op.cit., p. 405, and Grude, op.cit., p. 97.

[9] On the 'Lofthus uprising'–named after its dominant hero–cf. inter alia Koht 1926 ch. 22, Steen, *NFLH*, vol. VIi, pp. 104–124, and Georg Sverdrup, *Lofthusbevægelsen*, Kristiania 1917.

[10] Cf. Steen, op.cit., pp. 105 ff.

[11] Steen, op.cit., and Koht 1926, pp. 306 ff.

was exempted if it was invested in trade and seafaring instead of loans to the landholders. This might have stimulated the increasing claim for debt redemption.

These conditions led to peasant complaints about trade regulations and taxation from the mid-1780s. A Norwegian peasant, Christian *Lofthus,* who had been turned away in Copenhagen because he brought no signatures in addition to his own, started to mobilize the peasantry for collective petitions in the communities of southern Norway. The uprising spread when the leader was declared an outlaw, and after his imprisonment a captured bailiff had to be rescued by mercenary troops. The movement died away after the leader had been imprisoned for life and several others got severe sentences, while the Danish monopoly on corn import was eventually abandoned in 1788.[12]

This event shared some of the characteristics of the preceding one. It was a regional movement without national co-ordination, and it was rather epidemic with little endurance beyond the immediate grievances. But it also showed definite signs of a wider social assembling, and connected the acute complaints about intolerable conditions with a general resistance against the urban commercial class and the mercantilistic privileges which were enforced by the provincial bureaucracy.[13] There were also other instances of peasant unrest in Norway around the year 1800–particularly in opposition to the sawmill privileges (cf. chapter 5), but they remained scattered, unco-ordinated, epidemic. The link between the various districts and local communities was still basically monopolized in the hands of the provincial officialdom.

Around the turn of the century, however, the first nation-wide and sustained popular movement spread through the Norwegian countryside. That was a pietistic lay movement within the Lutheran church, which challenged the sprititual guardianship of the rationalistic clergy.[14] The revival by the Haugian movement–named after the founding preacher–was based on a fundamentalist approach to the Book, which definitely implies a democratic aspect by making the text directly accessible without the interpretative mediation of the learned priesthood. The revival also implied a personal religious commitment, disattached from the authority of ecclesiastical officialdom. There was thus a

[12] Steen, op.cit., pp. 90 ff., and Koht, op.cit., pp 326 ff.

[13] On the antagonism between peasantry and bourgeoisie in Norway, see also Koht, "Bonde mot borgar i nynorsk historie" (from 1910), reprinted in *Nye innhogg og utsyn*, op.cit., pp. 93–138.

[14] It has been written extensively about the preacher Hans Nielsen Hauge–who was particularly active from 1976 to 1804–and his movement. The implications for the peasant movement are explicitly discussed, inter alia, by Steen, "Hans Nielsen Hauge og bondereisingen", in *Tusen års norsk historie*, op.cit., pp. 109–121, and by Koht 1926, op.cit., ch. 23.

definite connection between the popular revival and a general opposition against the public authorities, although the preacher himself had no thought of a general peasant movement.[15] H. N. Hauge also organized communities of 'friends'–often from the higher strata of the peasantry–on his travels in the country, and economic activities like farming and local industries were systematically stimulated–on a cooperative basis at first, but later individually on loans or security from the 'friends'. The activities of the revival movement were perceived as a threat against the public authority, and the leaders were successively prosecuted and imprisoned from 1804.[16] The revival established, however, a nation-wide network of social connections outside the public control, and it is significant that the dominant peasant politicians at the national and local level several decades later were active Haugians.[17]

The profile of the political peasantry in the first couple of decades after 1814 is at first sight diffuse and contradictory. There is little doubt that the public bureaucracy in actual fact increased its power by the events of 1814. During the last phase of absolute monarchy, the civil servants had ruled Norway in the name of the King. By the constitutional reform they came to rule in the name of the people–and despite their potential influence at the polls, the people were fairly powerless.[18] The peasantry disposed of a clear majority of the votes for the *Storting,* but either they did not participate in the elections, or they voted for the local officials as a reflex of patronage (cf. chapter 8), or they were unable to unite in a consistent opposition against the trained superiority of the officials in the parliamentary debates.[19] More conspicuous still, the peasantry showed a notable suspicion–or even overt hostility–towards the new constitution and the *Storting* itself.[20] The *Storting* was–rightly so–regarded as a stronghold of the

[15] Steen, op.cit., pp. 110 f.

[16] Hauge was stopped by reference to the Conventicle Act–*konventikkelplakaten*–of 1741–which had been adopted to check the pietistic Moravian propaganda in the twin Kingdom, and which prohibited laymen from leading edifications without the consent and presence of the parish priest.

[17] Cf. Steen, op.cit., pp. 119 ff., and Koht 1926, op.cit., p. 352.

[18] This has been convincingly argued by Sverre Steen–see, inter alia, *Det frie Norge,* vol. I, op.cit.

[19] There were, however, some peasant initiatives in the first sessions of the *Storting*–inter alia, for the transformation of the remaining ecclesiastical property into peasant freeholds, as we saw in chapter 6. But there were no consistent and unified peasant politics during these first years. Relatively solitary peasant representatives on the *Storting* session of 1815–like Teis *Lundegaard* and Hans *Barlien*–showed some leadership ambitions, however. Barlein fought for instance consistently against commercial privileges, sawmill privileges, and unjust taxation, and for general conscription and sale of the ecclesiastical property to the leaseholders–cf. Koht, "Hans Barlien", *Norsk Biografisk Leksikon,* vol. I, Kristiania 1923.

[20] Cf. Steen, *Det frie Norge,* vol. II, Cappelen, Oslo 1953.

civil servants, a twin-bastion with the Cabinet, which further was rather expensive and imposed heavy burdens on the peasantry, without much recompense to their benefit. This anti-constitutional suspicion seems to have been most pronounced among the higher strata of the rural population, the relatively large freeholders in the eastern districts, and it culminated in a royalist, non-parliamentarian rising in the year 1818.[21] The peasantry were generally hostile towards any taxation and public duties, and their dissatisfaction increased spontaneously when the authorities imposed an extraordinary tax–the 'silver tax' (*sølv-skatten*)–which was intended to supply the financial foundation for a national Bank of Norway. The unrest is said to have been nurtured by a Swedish royal agent, *Röslein,* who travelled the country and told the peasantry that things would take a turn for the better if the power of the King was strengthened at the expense of the *Storting.*[22] The crop failures of the autumn 1817 further aggravated the situation. The peasantry mustered in several eastern districts in the spring of 1818, and sent a collective petition to the *Storting,* whereby they demanded a relief of their economic burdens, particularly by cheap loans for the taxes, and by the accomplishment of a monetary reform which implied that the Bank of Norway should issue money to be distributed as a compensation for the 'silver tax'.[23] The majority of the *Storting* refused to consider the petition, while the economic difficulties increased throughout the springtime.[24] The peasant unrest, with agitation and petitions to the *Storting* for lower taxes and easy loans, was now–at least partially–coordinated by the well-to-do freeholder Halvor *Hoel,* who represented a rationalistic and anti-clerical attitude far from the politically more tranquil Haugian movement.[24b] When a renewed petition, with signatures from the eastern areas, was refused by the *Storting* in August, the rural unrest grew sharply, and peasants from

21 See particularly Steen, op.cit., pp. 244–284. Cf. also Lars Reinton, *Folk og fortid i Hol,* vol. II, Oslo 1943, pp. 17–30, and Koht (ed.), *Fra den gamle bondeopposisjon. Brev og innlegg fra Stortingene 1818, 1821, 1822 og 1833 av Jacob Hoel,* Historisk Tidsskrift 2.–4. levering 1926, Oslo 1927, pp. 8–15.

22 Röslein was ordered by the viceregent to leave Norway in the spring of 1817, and it is dubious how far his mandate from the king, or crown prince Charles John, went.

23 The monetary reform had been proposed by an oppositional politician, war commissioner *Sebbelow,* who was allied with some of the peasant representatives in the Storting–cf. Steen, op.cit.

24 The timber market for the eastern area is supposed to have failed, and the weak public finances led to stronger claims for unpaid taxes–Steen, op.cit., pp. 255 ff.

24b Halvor Hoel, from the central eastern county of Hedmark, had been elected to the national assembly in 1814, but his admittance was refused with reference to a dubious dismissal of a criminal charge some years earlier. His anti-constitutional feelings were no secret, and he also retained contact with the expelled royal agent Röslein. His brother Jacob Hoel was an oppositional leader in the first parliamentary sessions after 1814–cf. Koht (ed.) 1927, op.cit.

several counties gathered and marched towards the capital. They repeated their former claims, but they also sent an expedition to Charles John—who was to be crowned as King of Norway later in the year—to offer him absolute power at the expense of the *Storting*. The crown prince refused, but a new wave of unrest and agitation spread during his coronation in September, and (possibly) thousands of unarmed peasants marched against the capital. They were surrounded by military troops outside the town, and several hundreds were arrested. The rising now faded away, while some of the leaders were sentenced after a commissionary investigation.[25]

The preceding account should be sufficiently detailed for an evaluation of the general nature of the events: The peasant risings of 1818 originated, at least in some of the particularly affected areas, from acute economic strain, where the freeholders perceived their ancestral farms in danger. The type of unrest was rather similar to the major risings under the absolute monarchy in the second half of the 18th century, by turning against the tax-collecting state and its representatives, and by being quite epidemic in character. But in addition it now had a clearer political aspect, by demanding the restoration of monarchic power to the neglect of the new constitution. Charles John should obviously take co-responsibility for the agitation which nurtured this demand, although he definitely drew back from the consequences.[26]

There was no substantial non-parliamentary peasant movement during the 1820s, and the peasant representatives to the parliamentary sessions were still unable to consolidate their group into a consistent body of concerted political action. Some common interests might have existed between the commercial class and segments of the peasantry as the major 'taxpaying classes' in opposition to the bureaucracy, but their parliamentary conduct was poorly unified, and the peasant group was definitely not able to hammer out a distinct political platform on its own.[27]

This strategical immobility changed rapidly from the early 1830s, and the next couple of decades came to constitute the classical period of the distinctive Peasant Politics in Norway. The early 1830s were still economically difficult for the agricultural population. Rising prices

[25] Halvor Hoel, who had not personally taken part in the march against Oslo—or Christiania as it was called then—was sentenced by the High Court of Justice several years later, but was virtually pardoned by the king.

[26] Cf. Steen, *Det frie Norge*, vol. II, op.cit., pp. 299 ff., and Reinton 1943, op.cit., pp. 29 f. See also W. Keilhau, *NFLH*, vol. VIII, pp. 218 ff., and B. A. Nissen, *Vårt folks historie*, vol. 6, pp. 196 ff.

[27] Cf. Steen, *Det frie Norge*, vol. III, Oslo 1954, pp. 120 f. and pp. 260 f., and ibid., vol. V, Oslo 1962, pp. 241 and passim.

on grain led to particular problems for the corn-deficit areas, while crop failures and aggravating trends of trade for timber affected other sections of the peasantry.[28] Some increase in the need of cash, together with a decrease in convertible goods, thus entailed a greater frequency of emergency loans subject to usury, since the banking system was still rudimentary (cf. chapter 7). The provincial officials were, as ever, obliged to enforce payment and manage bankrupt estates, so their popularity in the rural areas far from improved. The general anti-bureaucratic opinion of the time was also stimulated by overt propaganda. The peasant agitator John *Neergaard* issued his famous pamphlet *Olaboka* in 1830, and travelled the country for about two years, slightly disguised as a pedlar to avoid the accusation of subversive activities.[29] Neergaard particularly made contacts with the leaders of the Haugian movement (though not a Haugian himself), and his teaching was simple enough: to fight the burdensome bureaucratic supremacy by voting for peasant candidates to the *Storting*. This policy was supported by the importunate newspaper *Statsborgeren* ("The Citizen"), which particularly circulated among the larger freeholders in the east during the early 1830s.[30] The breakthrough for the peasantry at the *Storting* in 1833, when they doubled their representation and clearly outnumbered the civil servants, could hardly have been conceived without these activities (cf. chapter 8). The peasant movement now opted for the parliamentary potentialities opened up in 1814, and took an anchorage in strict constitutional defence.[31]

The peasant group was never disciplined for consistently joint action, but the basic features of their policy were effectively advocated by, among others, an outstanding peasant leader for several decades to come, Ole Gabriel *Ueland,* who represented a constituency in the relatively egalitarian south.[32] The basic ingredient in the national peasant politics during the 1830s and 1840s was to cut down the public expenses and reduce the items of expenditure. The program of economizing was primarily directed towards the state budget, and affected items like the salaries or pensions for the 'non-productive' civil servants, the outlays on higher education–including the scientific agricultural schools

[28] Cf. Ibid., vol. V, pp. 253 f.

[29] Cf. Koht, "John Neergaard", in *Norsk Biografisk Leksikon,* vl. X, Aschehoug, Oslo 1949; Koht, "Bondepolitikken", *Samtiden* 1908, reprinted in Koht 1964, op.cit., pp. 83 ff.; Steen, op.cit., pp. 251 ff.

[30] The first editor was Peder *Soelvold.* The paper ended already in 1837. Cf. Steen, inter alia, pp. 254 f. and pp. 340 ff.

[31] The bureaucracy, on the other hand, was no longer so keen on constitutional purism.

[32] Cf. Arne Bergsgård, *Ole Gabriel Ueland og bondepolitikken*, vol. I–II, Aschehoug, Oslo 1932. Cf. also the crucial review article by Steen, "Arne Bergsgård: O. G. Ueland og bondepolitikken", *Historisk Tidsskrift*, vol. 29, pp. 325–340, reprinted in Steen 1958, op.cit. See further Steen, vol. V, 1962, op.cit., pp. 271 ff. and pp. 321 ff.

(although important elements in the primary education system were supported provided they were not too expensive), or the means of social security reforms. This rather consistent attitude was supplemented by principles of the following kind:

- a quite extreme version of local self-government was to replace the system of centralized provincial administration. The peasant opposition thus opted for democracy on a decentralized, communalistic basis, since a high degree of *local* freedom would imply peasant control of public expenditure (cf. chapter 8).
- bureaucratic control–and the burdensome fees to the provincial officialdom–was to be replaced by elected commissioners. This demand closely related to the peasant conception of local autonomy.
- the peasant duty of traveller conveyance was to be suspended, since it was a degrading nuisance in the rural areas.
- the economic policy of the peasant opposition was slightly more complicated. They definitely–and successfully–opposed the general tax on registered land, which was nearly exclusively detrimental to the landholding peasantry. The public revenues should rather be supplied through high import duties, particularly on luxury goods, which were quite irrelevant to the peasantry at the time. They also generally favoured protective import duties on agricultural produce, but there was some discord between livestock-rearing and corn-producing areas.[33] The peasantry also advocated a policy of low rates of interest, to be regulated by the state, since they generally obtained little monetary surplus, but needed cheap money for inevitable obligations. Finally, they opposed all mercantilistic privileges on domestic trade and crafts, and were consistently liberalistic in this respect.[34]

Measured by conventional ideological categories, the classical peasant program is bewildering even in a conventionalized from. It is definitely 'liberal' in its opposition to commercial privileges and guild monopolies. But it is equally 'illiberal' so far as the policy for rates of interest and import duties are concerned. It can also hardly be labelled 'liberal' in the cultural sphere: the peasantry opposed general religious liberty in the 1840s. They certainly fought for abolition of the Conventicle Act (cf. note 16 above)–and succeeded in 1842–but they were highly restrictive against Jews and Christian dissenters outside the national church. The principle of religious liberty was thus used

[33] O. G. Ueland, for instance, represented a corn-deficit, cattle-breeding locality, and favoured a customs tariff on butter, while he advocated free import of grain.

[34] Cf., inter alia, Bergsgård 1932, op.cit., vol. I, pp. 196 ff., pp. 322 ff. and passim.: Steen, vol. V, 1962, op.cit., pp. 452 f.

as an argument only so far as was necessary for the 'low church' Haugian movement, which involved the old peasant struggle against bureaucratic supremacy.[35] The peasant program was 'democratic', but in a qualified sense. The question of extension of citizenship to the non-possessive classes met with sympathy in the structurally egalitarian south and west, while the larger eastern freeholders–facing an increasingly numerous class of crofters and rural proletarians–were definitely restrictive. Generally, the peasant emphasis on 'popular' government lay on communalistic autonomy, with decentralization as the primary ingredient. The peasantry were, however, rather 'conservative' in their general attitude towards the constitution of 1814, and equally so in their continuously royalist inclinations.[36]

This rather confused mixture of ideological ingredients demonstrates clearly that the labels as such are utterly misleading, and deserve reformulation in more adequate categories. The peasant program was in fact a sustained expression in defence of peasanthood and peasant autonomy. It seems modelled to defend a way of life rather than the scope of entrepreneurship in a societal order dominated by non-peasants. This is not, however, immediately obvious. There is, seemingly, an ambiguity in the peasant program. Several principles serve traditional communalistic protection against non-peasant influence. Others, like those on liberalization of trade, apparently point towards more expansive mobility in a wider economic sphere. Partly, of course, the program expresses a compromise between varying agricultural conditions–between the more corporate peasant communities producing for subsistence, on the one hand, and those closer to urban markets, more markedly integrated in commercial networks, on the other. But it expresses only partly a compromise, since there is yet no real contradiction between these principles. As we have argued earlier, no community was ever self-sufficient, and the market was an economically strategic field even for the subsistence-oriented peasantry. The early peasant politics were thus a pragmatical common denominator for the opposition against bourgeois privileges and bureaucratic power, where the state apparatus, with its provincial ramifications, appeared as a costly superstructure on an expansive commercial society where the peasantry were inferior.

The peasant politics thus correspond closely to those structural char-

[35] Cf. Aage Skullerud, *Bondeopposisjonen og religionsfriheten i 1840-årene*, Univ.forl., Oslo 1971. See also, for a slightly more diverting interpretation, Bergsgård 1932, vol. I, pp. 433 ff.

[36] Cf. Steen's review of Bergsgård 1932, op.cit., and Kaartvedt, "Politiske retninger 1814–1869", in *Det Norske Storting*, vol. I, passim. See also Keilhau, *NFLH*, vol. X, pp. 9 ff. and passim, for a rather puzzled and unappreciative account of the peasant program.

acteristics prevailing during the first half of the 19th century which were analysed in chapters 6–7 above: the relatively high degree of social cohesion among the landholding peasantry–prior to the more well-defined individualization of land and farm management–was combined with quite traditional methods of farming and a low degree of commercial and financial penetration. Transitional forces were simultaneously under way, and threatened an agrarian solidarity which –because of the notable structural heterogeneity–was based on distance: an active defence of localism was thereby a necessity. Along the functional dimension, this defence was fought on behalf of the primary economy in opposition to the secondary and tertiary sectors. Along the territorial line, it implied a defence of the principles of locality and household versus the pressure for integration by the state.[37] The ironical fact is that the *Storting* –as a clearly centripetal systemic factor–had to be used for such a purpose, although it was intentionally used to strengthen the peasant autonomy in local affairs.

The classical peasant politics nevertheless covered up latent contradictions which increased during the structural transformation of peasant society, and which became clearly visible around the middle of the century.

The political unity of the peasantry split open about 1850. It had increasingly been strained by heterogeneous agricultural conditions and uneven development of mercantilization. There was now a coincidence of antagonisms–between the common peasantry and the larger ones, between the more egalitarian southern and western communities and the stratified central and eastern counties–on the crofter system, the proletarian movement, and the customs policy.[38]

The crofter question had become politically acute. The number of crofters increased notably towards the middle of the century–particularly in the central (Trøndelag) and eastern areas (cf. chapter 6). Their conditions generally had taken a turn for the worse, and their circumstances were considered by the *Storting* in 1851. O. G. Ueland, and other peasant representatives from the southern and western regions, opted for greater freedom for the crofter class, like a legal abolition of labour duties in the crofter contracts. The peasantry from the eastern counties–with vested interests in the matter–were heavily against such compliance.[39]

This issue was pushed to extremes by a (mainly) rural proletarian

[37] Cf., inter alia, the general discussion of cleavage dimensions in Lipset & Rokkan 1967, op.cit., introduction.

[38] On the parliamentary debates on these issues by the middle of the century, see *Stortingets efterretninger*, 1836–54, III.

[39] Cf. Bergsgård 1932, vol. I, pp. 597 ff.; and Koht "Bondepolitikken", in Koht 1964, op.cit., p. 91.

movement which had originated two years earlier. There was, as we noted in chapter 7, a general economic depression from 1847 to the early 1850s. The crisis particularly affected the timber exports, and thereby hit those crofters and smallholders who were dependent on seasonal forest labour. It also entailed distraint farm sales in some rural areas. A radical proletarian movement, which sometimes engaged in rather vigorous demonstrations, started in 1849, and culminated in 1851 with nearly 300 local associations and a total of more than 20000 members.[40] The movement originated among the urban proletariate, but soon the crofters, agricultural labourers, and smallholders became numerically dominant, with mass support in the eastern counties north of Oslo, and in the central Trøndelag.[41] These were the areas where the rural socio-economic dominance of the large, consolidated freeholders –resting upon a labour force additional to family work–was most pronounced.

Even the *Thrane* movement showed some royalist inclinations in opposition to the *Storting* of the proprietor class; and its principal demands–expressed in a petition to the king in 1850–included the following main issues:[42]

- commercial liberty both in domestic and foreign trade, and particularly abolition of custom tariffs on corn and other necessities of life.
- improvement of the legal and material circumstances for the crofters.
- universal suffrage, general conscription, and improvement of the primary school system.

The propaganda of the movement was mainly directed against the larger peasant proprietors, and it thus–for the first time–forced the landholding peasantry to formulate an explicit attitude downwards, towards the landless, and not only towards the dominant urban world. The concerted voice of the lower classes–urban and rural alike–made

[40] The *Thranitter*-movement–after the leader Marcus *Thrane*–has been a relatively popular topic in historical research. The most informative monographs include O. A. Øverland, *Thraniterbevegelsen,* Kristiania 1903; Koht. *Tranitterbevægelsen,* Kristiania 1917; Jacob Friis, *Marcus Thrane,* Kristiania 1917; Oddvar Bjørklund, *Marcus Thrane–sosialistleder i et u-land,* 2. ed., Tiden, Oslo 1970; Rolf Grankvist, *Thranitterbevegelsen i Trøndelag,* Trondheim 1966. Cf. also an edition of Thrane's own speeches and articles, ed. by Einhard Lorenz, *Arbeidere, forén dere.,* Pax, Oslo 1969.

[41] Cf., inter alia, E. Lorenz, *Arbeiderbevegelsens historie 1789–1930,* Pax, Oslo 1972, pp. 26 ff., and Grankvist 1966, op.cit.

[42] Cf. Lorenz, ed. 1969, op.cit., pp. 8 f.

a breach in the old peasant politics, whereby the larger peasant proprietors in the east partially approached the governmental elites.[43]

This internal peasant conflict was reinforced by the question of import tariffs on bread grain. The tariffs were drastically reduced in 1851, by an alliance of peasant representatives from the southern and western livestock area together with the now liberalist officials, while the landholders of the corn-growing districts of central and eastern Norway voted for protectionism.[44] The issue had become acute by increasingly commercialized farming in the east, and revealed an open cleavage in the commodity market sphere (cf. chapter 7). The general peasant cleavage was, further, conditioned by the increasing social differentiation of the rural population, where the population growth still found no outlet outside agriculture. The proportion of crofters had increased particularly fast in the best areas of agriculture and forestry, where they came to constitute a crucial labour force for the new commercial activities of the larger peasant proprietors. The traditional patriarchal inclusion of the landless in the master household accordingly gave way to more purely economic and contractual relationships, which facilitated an independent political voice from below (cf. chapter 6). When it was heard, the established peasant unity split over the message.

The signs of a new political pattern, a coalition between the upper classes–urban and rural–against the threat from below, were not accomplished. It was a definite tendency under acute socio-political strain, but it was never consolidated. This is probably why the peculiar political system of a bureaucratic regime in Norway finally collapsed a few decades later. During the 1850s there were still vital issues which could rally the peasantry to joint action. The law on general conscription–for which they had fought continuously since 1814–was passed in 1854, and another part of the peasant program–the principle of equal inheritance for men and women–was also adopted during that session.[45] The crucial reform bill for development of the general education system got frank peasant support in 1860, although the peasant leaders had been reluctant–for financial reasons–towards a less epo-

[43] Cf. Koht, "Bondepolitikken", in Koht 1964, op.cit., p. 91, and Koht, "Bonde mot borgar . . .", ibid., p. 137. The peasant leader O. G. Ueland wrote that the labour movement had scared the eastern freeholders and transformed them from lions to hares (cf. Bergsgård 1932, I, p. 570), and he could support the demand for universal suffrage, although he dissociated himself from the free trade principle.
It should be noted that the *Thrane* movement was immobilized–and soon died away–by the prosecution of the leaders on the charge of subversive activities in the early 1850s.

[44] Cf. Bergsgård 1932, vol. I, pp. 601 ff. The abolition of grain customs was likewise, as we have seen, a leading issue for the proletarian movement.

[45] Cf. Bergsgård 1932, vol. II, pp. 31 ff. and pp. 69 ff.

chal proposition only twelve years earlier.[46] The restoration of the original 2 : 1 ratio between rural and urban representatives to the *Storting* – in 1859 – was a victory for the peasant opposition as a whole (cf. chapter 8). The peasantry also continued a unified fight against the remnants of commercial privileges, and succeeded gradually from the 1840s to the 1860s with support from bureaucratic laissez-faire ideology (which replaced mercantilistic doctrines during the first half of the century), against the commercial bourgeoisie (cf. chapter 7). The final crucial issue which should be noted, and which called for unanimous peasant action in several parliamentary sessions during the 1850s and early 1860s, was the defence of the allodial privilege, the *odelsrett* (cf. chapter 5). This peculiar institution, which had been intrinsic to the agrarian structure and peasant freehold of Norway for ages, had been criticized by bureaucratic and bourgeois interests since the 18th century, and governmental motions were carried for its abolition in successive parliamentary sessions from 1845 to 1861. The heavy peasant resistance was continually successful, so the institution basically survived.[47]

There were signs of a deepened cleavage between the peasantry and the bourgeois-bureaucratic elites again from the late 1850s. The *Storting* was now marked by an articulate group of urban radicals, particularly lawyers and newspapermen, who gradually approached the peasantry to rally them behind their democratic and nationalist aims.[48] There were also, as we saw in chapter 7, increasing economic problems from the late 1850s through the 1860s. The general depression from 1857 particularly hit the peasantry – as the major debtor class – with distraints and forced sales. The nation-wide peasant movement – as an explicit response to the prolonged crisis – developed rapidly in the late 1860s. The dominant peasant leader in the *Storting* – or perhaps particularly outside – was now Søren *Jaabæk,* who also represented the relatively egalitarian south. He issued the popular newspaper *Folketidende* from the summer of 1865, and advocated a ban on 'public waste' to fight the crisis. The peasantry, as the principal 'working class', was sup-

[46] Ibid., pp. 226 ff. and vol. I, pp. 316 ff.

[47] See Bergsgård 1932, vol. II, pp. 73 ff. Cf. *Stortingets Efterretninger*, 1836–54, III, op.cit., p. 460 and p. 837. It should be noted, however, that peasant defence of the *odelsrett* was not unanimous.

[48] This group was dominated by the lawyer Johan *Sverdrup*, who was to become leader of the successful constitutional opposition of peasants and urban liberals – in the liberal party of the Left, *Venstre* – against the regime in the 1880s, which eventually also brought Norway out of the union. The group tried to formalize its cooperation with peasant leaders in the (shortlived) *Reformforening* (Reform Society) at the parliamentary session in 1859, and also played some role through the equally informal *Collegium* during the sessions in the 1860s. Cf., inter alia, Koht, *Johan Svedrup,* vol. I, Aschehoug, Oslo 1918, pp. 280 ff.

posed to be impoverished by taxes and duties, and had consistently to oppose the supremacy of the privileged bureaucracy, which only led to intolerable tax burdens and indebtedness. This opposition had to be based on popular mobilization from below.[49] The organizational expression of the new peasant movement was to become 'The Society of the Friends of the Peasantry'–*Bondevennenes Selskap.*[50] The basic ingredients in the political program of the movement can be summarized as follows:[51]

- public expenditure had to be drastically reduced. There was here a vigorous revival of the peasant program from the 1830s, and S. Jaabæk now even opposed peasant liberal propositions of reform from the 1850s, like annual parliamentary sessions and a jury system in legal affairs, for financial reasons.
- liberalism in all kinds of trade and industry, not only for domestic trade and crafts, but also unauthorized professional work in medicine and law.
- protection of the debtor class against the plutocracy, by minimizing the need of money and fixing the rate of interest at the artifically low level of 4 %. The *Folketidende* accordingly agitated against the *Hypothekbank* (Cf. chapter 7), since financial credit was perceived as a necessary evil during periods of crisis, rather than as a productive factor.[52]
- democratic reforms like universal suffrage, direct elections to the *Storting,* and improved general education. While Jaabæk consistently opposed public expenditure on advanced studies, including the agricultural schools, he favoured general education as a progressive demand.

The basic elements of this program seem to express agricultural conditions which were far from universal–and which definitely were declining–in Norway during the late 1860s. The restrictive attitude to-

[49] Cf. *Folketidende,* no. 2, 19 July 1865.

[50] The main studies of this association–or its local branches–include Carl Lund, *Jaabæk og bondebevegelsen i Bratsberg amt,* (handwritten) thesis in history, University of Bergen 1926; Svein Jansen, *Bondevennbevegelsen i Norge 1865–1880,* thesis in history, University of Bergen 1965; Dagfinn Slettan, *Søren Jaabæk og bondevennbevegelsen i Lister og Mandals amt,* thesis in history, University of Trondheim 1972. Other studies are presently in course of preparation. Cf. also Koht, *Johan Sverdrup,* vol. I, pp. 402 ff., and Koht, "Søren Jaabæk", in *Norsk Biografisk Leksikon,* vol. VI, Oslo 1934.

[51] Some shift of emphasis occurred during the years from 1865 to the early 1870s, but these will not be analysed in detail here.

[52] Cf. Kaartvedt, *Kongeriket Norges Hypotekbank 1852–1952,* Part 1, op.cit., pp. 144–151.

wards the rate of interest, and generally the hostility towards all types of public expenditure, implied a protection of the peasant communities from the expansive world of money and credits. It must have been particularly attractive to those rural areas which still were only marginally dependent upon commercial agriculture. The propositions for extended democracy–by universal suffrage in particular–also found little response among the larger central and eastern freeholders, with their exploitation of a numerous–although declining–rural proletariate. The program was principally acceptable to the relatively egalitarian counties in the south and west. Still it included fairly adequate issues in defence of the rural lower classes even elsewhere.

This general interpretation seems confirmed by the diffusion of the movement. The local societies of *bondevenner* originated in Jaabæk's home parish in the south (Mandal) in 1865, with more than a hundred members. Then some new ones were established from the summer of 1867, while the first national conference took place in 1868. In 1870 the total number of members has been estimated at about 25 000, with the centres of gravity in the southern and western counties.[53] The movement was the first nation-wide voter organization, and it aimed explicitly at the *Storting* elections in 1868 and 1870. The 1868 election also became a remarkable success for the peasant candidates, who got a majority of the seats, and the campaign was continued in 1870 (cf. chapter 8).

It should be noted, however, that Jaabæk never got an undisputed leadership position among the peasant representatives in the *Storting*. His most reliable followers mainly represented the southern and western counties, as an expression of the old cleavage from the middle of the century.[54]

The peasant movement declined from the early 1870s. The boom period of the 1870s also entailed a notable agricultural improvement, which made political mobilization on communalistic lines less urgent.[55] The central leadership of the *bondevenn*-association probably also offended the bulk of the peasantry by increasingly anti-clerical and anti-royalist attitudes–Jaabæk himself was even a republican–and their demand of universal suffrage was either an irrelevant or a provocative issue among the peasantry.

So the prospects of a broad, popular, lower class movement, in-

[53] Cf. Slettan 1972, op.cit., pp. 15 ff.

[54] Cf. Rolf Danielsen, "Grupper og partier i Stortinget", *Det Norske Storting gjennom 150 år*, vol. II, pp. 335 ff. Jaabæk was also actually a more doctrinaire economic liberalist than Ueland or probably any other peasant representative–cf. Koht, "Søren Jaabæk", *NBL*, VI, op.cit., p. 564.

[55] Cf. Koht, *Johan Sverdrup*, vol. II, 1922, p. 58.

[56] Slettan 1972, op.cit., pp. 118 ff., makes a central point of this.

cluding the self-employed peasantry, eventually failed. The leader of the oppositional liberal intelligentsia, Johan *Sverdrup*, was increasingly able to rally the peasantry behind him in a consistent nationalist-democratic strategy against the governmental elite. Jaabæk had even offered to follow him, by a favourable proposal of agreement, in 1868–69.[57] The original antagonism between the two nevertheless continued occasionally during the early 1870s, but basically the new pattern implied peasant adherence to the liberal-nationalist trend aiming towards parliamentary power and national self-assertion. Far from communalistic autonomy, this policy intended to obtain central power through the state apparatus. Although the peasantry were still able to dissent on basic communalistic issues, they consistently adapted to a nationalist strategy which intended to exploit central power rather than to pare it down.[58] The bureaucratic elite and the patrician bourgeoisie accordingly leaned towards Sweden and the union as a protection against peasant power.[59]

The decisive end of the classical peasant politics, after a final convulsion in the late 1860s, is intrinsically concomitant to that structural transformation of peasant society which was analysed in Part Two.[60] The peasantry became economically linked up with the national networks, whereby mercantilization and individualized tenure drew them out of the local communities and created demands–for communications systems, technical knowledge, and credit supply–which could not be met on a purely communalistic basis. Dominant sections of the peasantry structurally gravitated towards the individualized and universal middle class, rather than towards those pre-industrial lower classes which the *bondevenn*-movement intended to embrace. Their

[57] He first offered to accept annual parliamentary sessions and the jury system, against support in the rate of interest issue and the abolition of traveller conveyance, but later even suspended his insistence on the fixed interest rate–cf. Koht, *Johan Sverdrup*, vol. I, pp. 444 ff. and pp. 457 ff.

[58] The qualifications of this account are outlined by Rolf Danielsen, *Det Norske Storting*, vol. II, op.cit. Cf. also Ingrid Semmingsen, "Ludvig Daae og de politiske gruppedannelser i 1870-årene", *Historisk Tidsskrift*, vol. 36 (1952–53), pp. 1–29, reprinted in *Hundre års historisk forskning*, op.cit., pp. 310–333. It should also be noted that the struggle for reconstruction of the Norwegian language–as a linguistic emancipation from Danish influence–met with little enthusiasm among the peasantry before the last quarter of the century.

[59] It might be noted, although it falls outside the scope of the present study, that after the breakdown of the bureaucratic regime–and parallel to increasing industrialization–the conservative centre of gravity was transferred to an upper business class, which changed the pro-state policy of their predecessors into a liberalist doctrine of non-intervention and restrictive public economy. So cards were redistributed along several lines before the end of the century.

[60] This point has, following Sverre Steen, most explicitly been argued by Alf Kaartvedt, 1956, op.cit., pp. 16 ff. Cf. also Knut Dahl Jacobsen 1964, op.cit.

negative resistance against the public authorities gradually gave way to a positive emphasis on governmental services. This interpretation of the new acceptability of public engagement–and the strategy towards exploitation of it–is, of course, no adequate or sufficient explanation of the *motives* behind the transformed peasant politics around 1870. It constitutes, however, a basic structural precondition for the political reorientation which definitely took place. The structural transformation in some respects also implied an equalization of agricultural conditions–firstly by gradually integrating the latecomers into the commercial world; secondly by the notable tendency of farm partitions, which counteracted the formation of a rural aristorcracy; and thirdly by making the larger farmers less marginally dependent upon their own family labour, since livestock husbandry became more generally adopted, while the number of crofters and hired landless steadily decreased (cf. chapters 6 and 7). The old structural cleavage between different peasant worlds thus declined, while individualization of tenure and increasingly entrepreneurial components in farming made the rural population as a whole more similar to the lower bourgeoisie, the middle classes, in general. The specific insistence on peasanthood had lost some of its actuality (we are not saying more than that), and the stage was set for political inclusion in a wider liberal movement.

Denmark

The basic characteristics of the peasant movement in Denmark are more easily surveyable, and its history before about 1870 is a far shorter one. The Danish peasantry were, as we saw in chapter 5, directly controlled–and highly subordinate–in their landed conditions until the final decade of the 18th century. Subject to adscription and villeinage, they were bound to the landlord and his estate not only economically, but personally, as legal subjects, as well. Unlike the situation in Norway, there was thus no unmediated confrontation between the peasantry on the one hand and bourgeois-bureaucratic elites on the other. The big landowners, with local jurisdiction over the peasantry on their estate, represented the landed interests within the wider societal order. Scattered outbursts of peasant protests occurred, but they were usually spontaneous assaults on a particularly brutal bailiff, and did not reach regional circles. The vertical segmentation of agrarian relationships–which effectively prevented concerted peasant action–implied landlord control of the horizontal linkages. The agrarian reforms of the late 18th century definitely meant a notable emancipation of the peasantry, and thus increased the potential of peasant mobility. But the peasants nevertheless remained principally passive for several decades to come, since the most radical type of articulation was scattered examples of joint petitions to the king and his cabinet about reduced leasehold charges and labour duties, or about improved opportunities for transition to copyhold or freehold (cf. chapter 6).

The first beginnings of an independent popular movement during the early decades of the 19th century were religious rather than political. The rural revivalist movements of the 1820s and 1830s should briefly be considered, since they obviously expressed a development which paved the way for the peasant politics in the subsequent decades.[61]

The traditional religiosity of village life had been collective and rather unconsciously handed down, with the church and the village priest as undisputed spiritual authorities. The individualizing dissolution of the village system, and the subsequent move from patrimonial landlordism towards freehold and purely economic obligations, also implied a cleavage between the peasant and the priest (cf. chapter 6). The pietistic popular revival correspondingly gave priority to personal sincerity about the faith, and organized religious fellowships and movements outside the traditional clerical dominance. The 'low church' revivals thus emphasized a development where the authority of the peasant was increasing, while the priest gradually lost his connection to the soil–literally the farm he had possessed as clerical officeholder–and his common interests with a village population. The peasant and the priest (a civil servant) thus ended up on opposite sides of the social struggle, with the peasantry left free to attack the burdensome tithe, while demanding that the governmental public clergy be replaced by a more truly congregational priesthood.[62] The theologically more broad-minded movement led by N. S. Grundtvig also became a crucial recruitment base for the peasant movement around the third quarter of the 19th century.[63]

The consultative assemblies of Estates became channels for an oppositional middle class movement from the early 1830s.[64] This was an opposition of urban liberals and the upper strata of the individualized, commercial agriculturalists against the conservative supremacy of higher civil servants, big landowners, and traditional guild monopolists. The majority of smaller peasants, dependent and semi-dependent, was

[61] This short account is based on P. G. Lindhardt, *Vækkelser og kirkelige retninger i Danmark*, Det Danske Forlag, Copenhagen 1951. See also H. P. Clausen, "Den sociale problemstilling ved udforskningen af de gudelige vækkelse", *Kirkehistoriske Samlinger,* 7th series, Vol. 6, No. 1, (1966), and Astrid Tønnesen, "Den sosiale baggrund for vækkelserne på Fyn i første halvdal af det 19. århundrede", *Kirkehistoriske Samlinger,* 1969, pp. 184–217.

[62] Lindhardt, op.cit., p. 26.

[63] Ibid., pp. 42 ff.

[64] Cf. Hans Jensen, vol. I, 1931, op.cit.; and Hovde 1943, op.cit., pp. 539 ff.

still largely passed by, and was not yet able to develop their own leadership.

The first popular peasant movement with a political program originated in the early 1840s.[65] The immediate occasion seems to have been the activity following the establishment of agricultural associations (*landboforeninger*), and incipient local government from 1841 (cf. chapter 8). The newspaper *Almuevennen* ('The Commoner'), edited by the craftsman J. A. Hansen, started in 1842 as an aspect of this activity, and it soon came to gain widespread influence among the peasantry on the islands. The principal background for the movement was, however, the increasingly acute antagonism between the leaseholding peasantry and the big landowners.[66] The agricultural boom from about 1830 (cf. chapter 7) implied that freehold became increasingly attractive to the peasantry, while leasehold, on the contrary, meant that the higher prices became capitalized to the benefit of the landlord. The economic prospects of landed property were exploited by higher entrance fees (*indfestning*) and leasehold duties. When the peasant movement gave this endemic antagonism a political form, the landlords counteracted by claiming abolition of the obligatory lease of 'peasant farms', in order to collect unified estates.[67]

The early peasant movement spread particularly on Zealand–where the percentage of leasehold was still fairly high, and thus the opposition to the big landowners most conspicuous. It was stimulated by the new associational life, by the new press, and, principally, by public meetings and personal propaganda. The dominant agitators in the early 1840s were the self-educated crofter Peder *Hansen*, and the school-teacher, of peasant origin, Rasmus *Sørensen*.[68] They drew considerable support for demands like general conscription, reduction of leasehold and labour duties, equal taxation on the *hartkorn* for privileged and unprivileged land, and fulfilment of the transition towards peasant freehold. A joint petition with such demands was delivered to the king in 1844, and it should be noted that the peasantry still predominantly favoured absolute monarchy, while the power of the provincial Estates was defended by the bourgeois liberals.

The growing unrest among the peasantry, with rather agitated popular meetings in several parts of Zealand, alarmed the landlords and the provincial officials, and the latter were instructed by the central authorities to follow the development closely. In the autumn of 1845, then,

[65] Cf. generally Kjeld Winding 1943, op.cit.; F. Skrubbeltrang 1938, op.cit., ch. VI; and Skrubbeltrang, vol. I, 1952, op.cit., pp. 111–144.

[66] Cf. Hans Jensen, vol. II, 1934, op.cit., pp. 522 ff.

[67] Cf. K. Hansen, ed., vol. I, 1925, op.cit., p. 194.

[68] Cf. the biographical accounts by Skrubbeltrang, vol. I, 1952, op.cit., pp. 112 ff. and pp. 117 ff.

the so-called 'Peasant Circular'–*Bondecirkulæret*–was issued, prohibiting public assemblies without permission and attendance by the local police.[69] The Peasant Circular expressed, provocatively, a public intervention in defence of landlord ownership, and far from appeasing the rural hinterland, it was met with fierce attack by the peasantry and the liberal press. A joint peasant deputation presented a petition to the king, while protests were noted from several rural areas.[70] The circular was then indulgently annulled so soon as in May 1846, but it had already been a mobilizing incentive to the peasant movement, which now became closely allied to the liberal bourgeois movement.[71] 'The Society of the Friends of the Peasantry'–*Bondevennernes Selskap,* also from May 1846–was the immediate organizational expression of this cooperation, and the executive committee was in fact completely dominated by the urban liberals.[72] While J. A. Hansen continued as editor of the newspaper *Almuevennen*, the rather royalist pioneers Peder Hansen and Rasmus Sørensen were largely passed by, as the leadership was taken over by more professional politicians. The peasant movement was now actually absorbed by the general liberal movement, which adopted the practical demands of the peasantry–like freehold on peasant land, freedom of trade, general conscription, equality in land taxation, and crofter reforms–against peasant support of the liberal struggle for a thorough constitutional abolition of absolutism.[73] The *bondevenn*-association engaged in canvassing before the elections to the provincial assemblies, and the movement got more than 3000 members in 1846, and more than 5600 by the summer of 1847–nearly exclusively on the islands.[74]

The bourgeois liberals were aptly called the 'national liberals', because complete inclusion of the duchies of southern Jutland, north of the river Ejder, with the kingdom of Denmark, was one of their principal political objectives. They increasingly alienated the peasantry by

69 See the comprehensive account by Hans Jensen, "Bondecirkulæret af 8. November 1845, Et bidrag til Kristian 8.s Tids Historie", *Historisk Tidsskrift,* 9.R. II, Copenhagen 1921–23, pp. 171–232.

70 Ibid., pp. 206 f.

71 Ibid., pp. 213 ff., p. 222, and pp. 230 ff.

72 Cf. Winding 1943, op.cit., pp. 15 f.; Skrubbeltrang, vol. I, 1952, p. 137; K. Hansen, ed., vol. I, p. 191 and vol. V, pp. 270 ff. The Norwegian *bondevenn*–movement of 1865, which was named after the Danish one, was, on the contrary, purely peasant-based.

73 Cf. Hans Jensen, vol. II, 1934, op.cit., pp. 527 f.

74 Ibid., p. 528.
The movement was particularly strong in the middle and north-western parts of Zealand, where leasehold was still widespread and landlordism dominant, while the following was less enthusiastic in northern Zealand, with more Crown land, and on Jutland, with more consistent peasant proprietorship–cf. *Den Danske Rigsdag*, vol. 3, op.cit., pp. 11 f.

concentrating on the national question around the year 1848, when the first war on Slesvig broke out against the Germans. The cleavage between the peasant movement and the national liberals seems to have stimulated radicalized peasant demands, still particularly centred on the islands, where the most outspoken rural inequality prevailed.[75] Peasant leaders like J. A. Hansen broke with the national liberals and favoured peasant candidates at the elections to the constituent national assembly in 1849. The liberals were often defeated, although the election turnout was quite low (cf. chapter 8).

There were, roughly, three major groups in the constituent assembly and the subsequent sessions of the *Rigsdag:* the conservative right based on the big landowners and the higher officialdom; the peasantry as a left-wing formation; and the national liberals as a centre group. During the early 1850s, however, there was still some co-operation between the peasantry and the liberal centre.[76] The cleavage deepened after 1854, when the national liberals formed their first Government, and from about 1860 there was a consistent strife between the two parliamentary camps. Although the peasantry had met with success on some of their demands–like general conscription (from 1848–49), and abolition of villeinage, the *hoveri* (from 1850)–they did not yet achieve obligatory abolition of the leasehold system, nor equal and universal suffrage for the local councils.[77]

In 1864 the national politics of the liberals definitively collapsed with the defeat in the second war on the duchies. The liberal centre was absorbed in a conservative restoration led by the landed upper class, the higher officialdom, and the bourgeoisie. This backlash revised the liberal constitution in 1866, and instituted the Upper Chamber as a guarantee against peasant power (cf. chapter 8). The peasant leader J. A. Hansen had even engaged in a constitutional compromise with the landlords against the bourgeois representatives.[78] His leadership position was, accordingly, weakened among the peasantry, and later the peasant demands of obligatory leasehold abolition and a democratic local election act were rejected jointly by the landowners and the bourgeoisie. This latter co-operation, confronting the peasantry, was

[75] Cf. Winding 1943, op.cit., pp. 17 f., and Hovde 1943, op.cit., pp. 539 ff.

[76] Cf. Holger Hjelholt, "Under Junigrundloven (1849–1866)", *Den Danske Rigsdag*, vol. 1, pp. 95 ff.

[77] Cf. Ibid., pp. 120 ff., and Winding 1943, op.cit., p. 23. The peasant program was opposed not only by bourgeois resistance against rural predominance, but notably also by some lack of peasant solidarity.

[78] This strange and shortlived cooperation between the landed aristocrat *Frijs* and J. A. Hansen was known as the 'October Association'–*Oktoberforeningen*–after a banquet in October 1865. The 'compromise'–to the satisfaction of the bourgeoisie–led to Hansen's acceptance of an upper chamber in the *Riksdag*. Cf. Povl Engelstoft, "Under Grundloven af 1866", *Den Danske Rigsdag*, vol. 2, pp. 21 ff.

confirmed by a mixed bourgeois-landlord Government in 1870.[79] The consolidation of the capitalist aristocracy–rural and urban–in a unified political Right, implied the final failure of co-operative peasant compliance. The peasant response was to gather their various groups in a unified party of the left–*Det forenede Venstre*–also from 1870, with a new generation of leaders imprinted by the liberal-democratic 'folk-high-school'-movement.[80]

The peasant party consistently combined their constitutionally democratic program with economic demands like abolition of leasehold and tithe, and was increasingly able to mobilize support among the rural population all over the country, including the freeholders of Jutland as well as the landless crofters.[81] This political mobilization is also reflected in the development of election turnout, which is documented in the previous chapter.

The original peasant movement in Denmark should be seen against the dual background of dominant landlordism and a relatively advanced stage of agricultural mercantilization towards the middle of the 19th century. Denmark had a rather substantial urban market, and agriculture also already produced extensively for export during the 1830s and 1840s (cf. chapter 7). The politically active sections of the peasantry thus made a strong bid for citizenship and taxation systems on equal terms with other classes within the national order.[82] The peasant movement was, above all, stimulated by a consistent demand for freehold in the still predominantly leasehold areas.

The Danish peasant movement was never, unlike the Norwegian one, genuinely communalistic. While the original peasant politics in Norway expressed freehold communities with a relatively low degree of individualized market integration, the Danish movement rather articulated an individualized market orientation which was immobilized by landlord ownership. The concomitant stress on elementary demands for equality and economic independence was at first instrumental to the aims of the liberal bourgeoisie, in a joint opposition to the interests of

[79] Cf. Ibid., pp. 57 ff.; Winding 1943, p. 27.

[80] Cf., inter alia, Skrubbeltrang 1938, op.cit., pp. 176 ff., and Winding 1943, pp. 29 ff.

[81] Cf. inter alia, Winding, op.cit., p. 33.
The particular peasant politics which had gradually gained some strength on eastern Jutland during the preceding couple of decades, were marked by a deeper antagonism towards the national liberals in Copenhagen and their anti-German policy, rather than towards the conservative landlords and higher officials. In the 1870s, however, this distinction was no longer relevant, while the peasant movement on the islands no longer had a repulsive bourgeois senior partner. Cf. Winding, "Valgkredserne gennem tiderne", *Den Danske Rigsdag*, vol. 3, op.cit., pp. 12 f.

[82] Their commercial orientation is probably also expressed by the increasing rural tendency to build credit institutions like insurance companies and taking over savings banks by giving them a peasant leadership, after the middle of the century. Cf. Skrubbeltrang 1938, p. 171.

landlordism and higher officialdom. Unlike the situation in Norway, the peasantry did not represent agriculture as a whole, but was rather agriculturally subordinate. Their political articulation was, accordingly, no localistic defence of landholding peasanthood, but expressed elementary social demands which might well be generalized in a wider liberal movement. The general rural antagonism towards the big landowners further conditioned a broader coalition downwards, including the landless within the peasant movement. This joint action of the landed inferiors entailed demands which eventually alienated the urban bourgeoisie, and pointed towards a consolidated peasantry facing the rural and urban upper classes. As suggested in chapter 7, the peasant proprietors successfully fought even the economic dominance of the landlords through the co-operative movement from the 1870s and 1880s, as an economic counterpart to the constitutional struggles in the political sphere.[83]

Sweden

The 'peasant movement' of late 18th-century and early to middle 19th-century Sweden is hardly traceable in a clear sequence. It is, in fact, hardly traceable at all, for reasons which will be given below. Those features of rural mobilization which–on different models–are found in Norway and Denmark, were virtually non-existent, while the Swedish peasantry, on the other hand, were already granted a legitimate voice in public affairs, continuously from the early 18th century. The latter fact may in itself, of course, have conditioned some lack of accumulated dissent.

Looking at the early period of the four-tier Estate Assembly, during the 18th century, the Peasantry were, as noted earlier, represented by the holders of crown land and taxable land, to the exclusion of the *frälsebönder* on privileged noble lands. The political role of the Peasant Estate was, however, subordinate and rather modest. The peasantry were generally royalist and alien to the nobility. They represented a support to the early bourgeoisie in its struggle against noble privileges, but were also clearly influenced by the clergy in their local constituencies as well as during the parliamentary sessions.[84]

The formal participation of the peasantry in parliamentary decision-making had definitely an improtant co-optational aspect, since the rep-

[83] The constellation from 1870 was then basically relevant until the political arrival of the industrial proletariate towards the end of the century. This phase is thoroughly analysed by Lafferty 1971, op.cit.

[84] Cf. Ragnar Olsson 1926, op.cit., pp. 111 ff.

resentatives then became responsible for taxation and other potentially controversial issues within their local communities.

Petitions and complaints from the rural population were brought forward at the *Riksdag* sessions after having been collected by the local public governors, so these petitions did not articulate any organized dissatisfaction beyond governmental control.[85]

The party cleavages on the *Riksdag* sessions during the second half of the 18th century are, moreover, of limited interest in an analysis of peasant politics. The *Hats* and the *Caps* were purely parliamentary parties, whose contest was based partly upon different views in foreign policy, and partly upon more general attitudes towards the constitution and the monarchy.[86] The peasant representatives were not consistently affiliated to either party, and there was even an interchange between the parties over time as to the most heavily royalist inclinations.[87] The peasantry, however, supported the nearly absolute Gustavian monarchy of 1789, in order to derive advantages from the King against the noble interests.[88]

The peasantry were primarily interested in the principles for disposal of land, and thus also wanted a guaranteed order of succession by lease of crown land, and a general equalization of the privileges on different types of land.[89] The peasant representatives further opposed the rather extreme method of field consolidation and reallotment during the first decade of the 19th century, the *enskifte,* which was conceived as a landlord pretext for acquisition of peasant land, and they successfully obtained less rigorous provisions, the *laga skifte*, a couple of decades later (cf. chapter 6).

The peasant politics were hardly based on any popular movement in the rural areas. The instances of non-parliamentary peasant unrest were few and rather peculiar in character. The most notable one had been the rising in Dalecarlia in 1743.[90] This event is, however, relevant to our topic. Firstly, because it was directed against conscription, which was to become a rather endemic theme in peasant complaints

[85] Ibid., pp. 119 ff.

[86] Cf. Gunnar Olsson, *Hattar och mössor. Studier över partiväsendet i Sverige 1751–1762*, Akademiförlaget, Gothenburg 1963. Cf. also Hovde 1943, op.cit., pp. 180 ff.

[87] While the *Hats* at first represented the nobility's defence of the 'free' constitution, they later became sceptical because of external warfare and internal political struggle, and then favoured monarchic power against the non-noble Estates–cf. G. Olsson 1963, op.cit., p. 241, pp. 296 ff. and passim.

[88] This concerned issues like the transformation of crown land into freehold, and confirmation of general peasant rights towards the privileged landowners–cf. chapter 6 above.

[89] Cf. Ingers, vol. II, op.cit., pp. 369–401.

[90] Cf. ibid., pp. 43–60. See also Bjarne Beckman, *Daluoppröret 1743 och andra samtida rörelser inom allmogen och bondeståndet,* Sth. 1930.

for more than a century; and secondly, because it expressed a general distrust towards a *Riksdag* which not only engaged the peasantry in warfare abroad, but which also lacked a stable paternal authority during an agonizing cleavage on the royal order of succession. When the Dalesmen refused conscription, and marched towards Stockholm in June 1743, they were roughly beaten down by armed forces, so the instance came to represent the last major peasant uprising in Sweden.[91]

The exceedingly dependent *frälsebönder* in the south, who fell outside the Peasant Estate and thus lacked all rights of political representation, were generally rather unresisting, but not continually so. There were notable movements of protest in the southern provinces of Scania and Halland during the early 1770s.[92] These provinces had, as noted in chapter 5, a great proportion of 'supreme noble lands'–*ypperligt frälse,* and the unrest particularly affected the peasantry under the manorial estates. The duties of labour and conveyance were imposed on them at the sovereign discretion of the landlords, and the peasantry appealed to the king, in order to get fixed regulations of labour duties and generally improved circumstances.[93] They also engaged in protest gatherings and labour strikes on several estates, and they even took legal proceedings against the landlords–although the verdicts of the courts invariably favoured the latter.[94] The movement was, on the contrary, suppressed, and several participants were given notice to quit their land.

The Peasant Estate in the *Riksdag* did not, as we have noted earlier, consider the problem of the *frälsebönder*, and their case was never included in the parliamentary petitions, the *besvär,* from the rural districts. Their fate was not, as a marginal phenomenon, put on the agenda for national peasant politics. The originally Danish institution of villeinage, the *hoveri,* thus prevailed several decades longer in Scania than it could be defended in Denmark itself, and the last protest movement among the Scanian *frälsebönder* occurred as late as the 1860s, when some prospects of freehold were in sight.[95]

Among the dominant sections of the Swedish peasantry, there was no real popular movement during the period from 1809 to 1866 either.

[91] The warfare in question was an armed dispute with Russia on the eastern province of Finland. The circumstances of the soldiers were said to be miserable, and Dalecarlia itself also suffered economic hardship.

[92] Cf. Staffan Smedberg, *Frälsebonderörelser i Halland och Skåne 1772–76*, Uppsala 1972.

[93] Ibid., pp. 125 ff.
The unrest seems to have been stimulated by agriculturally bad years, which particularly hit the peasantry.

[94] Ibid., p. 48, p. 55, p. 67.

[95] Ibid., pp. 190 f. Cf. Ingers, vol. III, op.cit., p. 311.

But the peasant representatives in the *Riksdag* increasingly articulated consistent peasant demands, for which they definitely were supported by the development in their rural constituencies. The peasantry were also able to produce quite outstanding political leaders on their own.[96] The basic characteristics of Swedish peasant politics before the representative reorganization in 1866 might be illuminated by an outline of the dominant issues. These were, roughly, the question of noble privileges, the system of conscription, the principles of economic policy, and the question of political representation. We will successively comment shortly upon them.

As regards the agrarian privileges for the nobility, the peasant demand for access, in principle, to noble land, even *ypperligt frälse,* met with success during the early decades of the century (cf. chapter 6). During this early period, the larger 'peasants of standing'– the *herrebönder*, who already possessed privileged land–were politically most aggressive among the peasantry. They were the better educated in a period when the general educational standard was low, and they were socially closer to the nobility, and thus more actually restrained by lack of formal possibilities.[97] The nobility, however, successfully refused the peasant demand for equalization of taxes on privileged and unprivileged land–i.e. an abolishment of landed privileges–while the peasantry capitulated, and virtually lost the battle for three-quarters of the century.[98] The peasantry also opposed the discriminating system of conscription, together with the obligation of providing troops, and the *indelningsverk,* which was a particular type of land allocation whereby the army–and parts of the civil and clerical officialdom as well–received land, provisions, and houses as remuneration directly from the taxable peasantry. These permanent peasant grievances were abolished through the introduction of general conscription and military reorganization in the 1870s.[99]

The peasantry also favoured freedom of domestic trade against the

[96] The main source for parliamentary peasant politics is of course the protocols of the Peasant Estate–*Bondeståndets protokoll*–which comprise more than a hundred volumes from the seventeen *Riksdag* sessions during the period 1809–1866 only. For the type of analysis undertaken here, it would have been rather an exaggeration to scrutinize them.

[97] Cf. Carlsson, "Bonden som politiker 1809_1866", in Ingers, vol. III, op.cit., pp. 238 ff.
The freeholders of the north were rather unconcerned about the question of privileges.

[98] Ibid., pp. 249 ff.

[99] Ibid., pp. 271 ff.
These issues had continuously been on the peasant agenda in the *Riksdag.* In 1811 there had even been an instance of popular uprising against extraordinary conscription –mainly among rural labourers and dependents in the south–where several participants had been killed by armed forces.

old mercantilistic privileges, and they seem to have given increasing priority to this demand towards the middle of the century. While the principle of restrictive financial economizing found widespread backing among the peasant representatives in the early decades of the century, this was less important to the *herrebönder*, and also to the liberalist phalanx that dominated peasant politics from the 1830s.[100]

Finally, the question of a reform in the system of political representation was given greater priority as the peasantry expanded on the landed property market.[101] They were no longer content with a fourth of the Assembly of Estates, and opted for joint elections to a new *Riksdag,* with the franchise restricted by qualifications of property or income.[102] This reform had equally, however, been the demand of the rising commercial and industrial bourgeoisie, and it was they, in particular, who had the greater share in the constitutional reorganization of 1866, whereby the upper chamber was explicitly designed as a defence against the numerical dominance of the peasantry.

The peasant representatives became, as we saw in chapter 8, quite dominant in the lower chamber during the first sessions of the new *Riksdag*. They were even able to unify their various groups in the Farmers' Party–*Lantmannapartiet*–from 1867.[103] This was a purely parliamentary party, without a national organization, and it was able to rally a majority behind most of its central issues in the second chamber.[104] These issues principally included:

- general conscription and reform of the militia system. The party generally favoured economizing in the public finances, which specifically affected its attitude to military defence.
- abolition of the traditional rural transport system, which implied peasant conveyance of travellers.
- reform of the taxation system–in particular removal of the discriminating land taxes.
- opposition to the principle of free trade, which had been adopted since the Crimean war. The party represented a moderately protectionist policy, on a compromise which partly and temporarily covered up the cleavage between dissimilar rural areas.[105]

[100] This also affected the peasant policy towards educational reforms–cf. ibid., pp. pp. 285 ff.

[101] See chapter 7 above. Cf. Martinius, *Agrar kapitalbildning*, 1970, op.cit., pp. 156. f.

[102] Cf. Carlsson in Ingers, III, op.cit., pp. 293 ff. Petitions to the Government advocating such reforms had received tens of thousands of peasant signatures in 1860 and 1865 –cf. Wallin 1961, op.cit., pp. 104 ff.

[103] The classical work is Edvard Thermaenius, *Lantmannapartiet. Dess uppkomst, organisation och tidigare utveckling*, Uppsala 1928.

[104] While it basically represented the interests of the peasant representatives, it also included a small, but quite influential, group of oppositional 'persons of rank'. Cf. Carlsson & Rosén, *Svensk Historia*, vol. II, op.cit., pp. 535 ff.

The Farmer's Party in the second chamber was mainly opposed to a conservative group of urban bourgeoisie and officials, who supported the political establishment, and who organized as the 'Intelligence Party' about 1868, and later reorganized as the 'Centre Party' from 1873.[106] In the late 1860s, however, the Farmers' Party also faced a radical challenge from the New Liberal Party, which was led by urban radicals from the professional middle classes, and which proposed liberal-radical reforms like universal suffrage for all men and women. This was definitely a weaker group than the corresponding ones in Denmark or Norway, with a less strategical leadership (the party disappeared already in 1871), but it had some success in attracting segments of the peasantry by vigorous agitation at non-parliamentary popular meetings, in support of peasant demands like abolition of the land tax system.[107] The politically most influential period of a consolidated peasantry was nevertheless based on the Farmers' Party from the early 1870s, until the cleavage on free trade in the late 1880s. The opposition between the individualized freeholding peasantry and the 'persons of rank'–the landlords in particular–dominated the political scene, and although the latent antagonism between the peasant proprietors and the numerous rural landless was socially crucial, it was politically nearly irrelevant during the period investigated here.[108]

This rough outline gives some clues to the basic pattern of peasant politics in Sweden. The transformation from a rather subordinate acquiescence during the 18th century to a more vigorous political articulation in the *Riksdag* during the first half of the 19th century expresses the gradually increasing relevance of wider socio-economic networks, without those dramatic counterpoints which characterized the development in Norway or Denmark. The pre-industrial Swedish society was marked by a privileged landowning nobility, which also formed the backbone of the public bureaucracy. The peasantry, of which the majority was quite independent, were principally engaged in the hallmarks of a social order based on assignment of privileges by Estates. Their demands comprised liberal reforms aiming at legal equality with their landed superiors, focusing on the questions of privileges, tax systems, conscription, and political representation. Parallel to the increasingly commercialized agriculture, they also demanded access to the market, in opposition to the mercantilistic monopolies

105 Thermaenius 1928, op.cit., pp. 66 ff and pp. 203 ff. The *Lantmannaparti* broke up over customs policy in 1888.

106 Cf. Carlsson & Rosén, vol. II, op.cit., p. 537; and Wallin 1961, op.cit., p. 440.

107 Cf. Wallin 1961, op.cit., pp. 154 ff.; Thermaenius, ed. 1931, op.cit., p. 9.

108 Thermaenius 1928, op.cit., pp. 91 ff.
To some extent, this latter conflict also eased off, or became less acute than it would have been, owing to mass emigration–cf. chapter 7 above.

of trade and crafts. The character of these demands, however, was also crucial to the rising commercial and industrial bourgeoisie. The dissolution of strict boundaries between the Estates had taken place within the agrarian framework, but it was politically fulfilled in dual action by the peasantry and the bourgeoisie, against the landed aristocracy–partly parallel through their respective Estates, and partly in concerted action under bourgeois leadership. Formally, the peasantry did not gain much from the bourgeois constitutional reform of 1866, but they were–for some time–able to unify their groups in a political party which challenged the supremacy of the lords, while the latter gradually fused with the higher strata of the bourgeois world.[109]

Peasant politics and patterns of alliance–a comparative summary

We are now able to identify the features of the Scandinavian peasant movements in relation to the potential scope of politics which was outlined in the last section of chapter 4. Our intention here is mainly a rough comparative summary, while the major qualifications are found in the preceding account.

The peasant politics in *Norway* might fruitfully be divided into four phases from the late 18th century to about 1870. During the first phase, prior to the distinctive articulation of political peasanthood from the early 1830s, the basic peasant drive had a spontaneous *ad hoc* character, with rather few signs of long-term coordination. If the peasantry acted politically prior to 1814, it seems to have been rather expressively, within localistic, or at most regional, dimensions. If their action was supra-regional and sustained, on the contrary, it was rather of a religious and non-political kind. This pattern was not immediately transformed with the constitution of 1814, although peasant unrest in the subsequent years probably had some elements of guidance from outside, owing to the greater relevance (potentially) of the peasantry in national politics.

The classical period of independent and exclusive peasant politics, during the 1830s and 1840s, was marked by a strategical exploitation of that central political influence which was implicit in the constitution –basically in defence of the localistic particularity of peasanthood

[109] It should also be noted that the 'bourgeoisie' was not–as little as the nobility–a uniform entity throughout the period. The bourgeoisie was gradually transformed during the 19th century as it acquired an increasingly commercial and industrial character, while the Burgher Estate in the traditional sense was quite insignificant around 1870. The stratification of the bourgeoisie accordingly followed new dimensions, more intimately connected to proprietorship and financial standing. Cf. also Sten Carlsson 1949, op.cit., pp. 305 ff.

against supra-local authority and economic claims from outside. The major modification here, of course, is that rural participation was not remarkably high, either in national or in local elections, together with a continued–if declining–tendency to vote for non-peasant candidates. This fact modifies the *intensity* of conspicuous cleavages, more than the *relevance* of the political delineations.

The crucial difference during the 1850s and 1860s then was the pronounced tendency towards a cleavage within the peasant movement itself, principally based on the uneven development of mercantilization and rural differentiation.

Finally, there was a notable reorientation from the early 1870s, when the politicalized peasantry increasingly fused with the pro-liberal middle class, under the leadership of urban professionals, and adopted basic issues from the wider national-democratic movement. This reorientation occurred more or less parallel to the individualized response of mass emigration, when substantial sections of the rural lower class voted with their feet. The dissolution of recognized codes of communal action, during population pressure and penetration of the hinterland by commodity relations, might have been conditional to this dual occurrence.

The peasant movement in *Denmark* prior to the early 1840s was principally non-political. It was based on a popular revivalist counterculture in opposition to the rationalistic preaching of the authoritative clergy. The politicalized drive from the early 1840s soon became guided in close alliance with the liberal-minded bourgeoisie, while the gradual secession from this strategical embrace, during the 1850s, led to a fairly exclusive peasant politics at the national level about 1870, with support from the crofters in joint opposition to the urban and rural upper classes.

Although the Peasant Estate in *Sweden* had a formal political status from the early 18th century, there was no sustained peasant movement behind it, and their position in the Assembly was rather modest. The peasantry obviously opposed the disadvantages of landlord superiority, but they did so principally by consistent royalism and support of the other non-noble Estates. Among the dependent *frälsebönder*, on the other hand, there were some few epidemic outbursts of independent socio-economic protest, which, however, were never put on the agenda by the Peasant Estate. The peasant politics during the 19th century was basically confined to the parliamentary arena, where there was an increasingly strategical opposition to the restraints which followed noble privileges and Estate boundaries, concomitant with improving economic conditions for the freeholding peasantry. The peasant demands did mainly coincide with those of the rising bourgeoisie, although the constitutional reorganization of 1866 stimulated a peasant consolidation in the lower chamber, while the landlords increasingly fused with the higher bourgeoisie in the upper house.

This exceedingly rough outline thus shows that there was no unilinear sequence in the Scandinavian peasant politics, but an irregular, and sometimes contrasting, pattern of development. This should be accounted for by the dissimilar character of peasant agriculture, and the dissimilar types of linkages between the peasantry and the wider social order: The economic intergration of the Scandinavian peasantry took place in strikingly different agrarian contexts. In *Norway* the lack of landed lords implied an unmediated encounter with the bourgeois-bureaucratic authority. When there still was a high degree of semi-autonomous peasanthood to defend, there were actually no relevant coalition partners. The exclusiveness of the classical peasant politics in Norway is thus closely in accordance with the structural composition of agrarian relationships; firstly as the lack of a semi-feudal landed aristocracy made an early peasant-bourgeois alliance unfeasible, and secondly as the peasant character of agriculture as a whole underlined the socio-economic peculiarity of the rural world. The cleavage occurred when rural differentiation became a socio-political threat to the upper agrarian strata, while commodity relations unevenly penetrated the countryside. The predominant opposition to the state-building upper classes nevertheless prevailed, and a coalition with the nationalist liberal-minded intellingentsia became increasingly acceptable with the transformation of peasant agriculture itself, since the corporate community was no longer a viable entity in opposition to the outside order. The pro-liberal urban leadership of this consolidated peasant opposition had been able to link the economic interests of the peasantry to its own national-democratic demands, and it modified the radical suffrage policy to satisfy the wealthier eastern farmers. The possible alternatives—either a centre coalition of the urban and rural middle classes, isolating the more radical group of smallholders and landless, or a fusion between the wealthier farmers and the conservative establishment—were thus both strategically played out at the political level.[110]

The *Danish* peasantry were immediately confronted with the landlords, who originally performed the economic and political mediation between the villages and the national society. The liberal-minded urban bourgeoisie likewise opposed the interests of the great landowners, the

[110] Both these alternatives in fact had supporters in influential political circles from about 1870. On the first—see Ingrid Semmingsen, "Ludvig Daae og de politiske gruppedannelser i 1870-årene", op.cit., and J. A. Seip, *Ole Jacob Broch og hans samtid*, Gyldendal, Oslo 1971, pp. 452 ff. On the second—see Anne-Lise Seip, "To-kammerhøyre foran forfatningskampen i 1870-årene", *Historisk Tidsskrift,* No. 2, 1973, pp. 138–157.

higher bureaucracy, and the traditional mercantilistic monopolists, and was at first able to rally a peasant mass basis behind them, since there was no peripherally autonomous peasanthood to be defended against the outside world. Substantial sections of the smaller peasantry shared crucial grievances with the crofter class–in common opposition to subordinacy on the land–and pressed for radical-democratic demands which gradually alienated the bourgeoisie.

Even in *Sweden* there was a joint peasant-bourgeois opposition to the privileges of the nobility, which was increasingly articulated as the peasantry economically approached 'bourgeois' status as independent proprietor-entrepreneurs. This politically dominant section of the peasantry was not, unlike the Danish one, directly dominated on the land. They shared the grievances of the rising middle class rather than those of the landless, whom they to a major extent exploited themselves, so the latter became hardly politically relevant during the period in question.[111] The consolidation of peasant politics at the parliamentary level occurred as the upper bourgeoisie and the nobles fused against them, while the professional pro-liberal middle class was too insignificant to play a consistently leading role in the opposition. By then, however, the 'peasant' characteristics of the agrarian population had been increasingly obliterated for several decades.

[111] While the rural proletariate in Norway faced an upper section of the peasantry which did not *dominate* the rural scene, since the more egalitarian areas were politically involved, and while the Danish crofters had some basic interests in common with the peasantry, the Swedish landless were virtually subdued by a hierarchy of superiors.

Conclusion

The conclusions that emerge from the present investigation should be introduced by a rough summary of basic agrarian features.

The *Norwegian* agrarian society was characterized by a personally and legally free peasantry–even under tenancy conditions. An initial and comprehensive move towards peasant proprietorship was already taking place from the late 17th century, and there was no village system regulated by landed superiors. Peasant freehold was cemented by kinship allocation of land through the allodial provilege, the *odelsrett,* and the commercially dynamic elements were to be found outside agriculture itself–in forestry, fishing, shipping or mining.

There was an increasing rate of population growth particularly from the early 19th century, with few outlets outside agriculture, and the number of crofters and rural labourers increased disproportionately towards the middle of that century.

The development of neo-technical farming (i.e. crop rotation, managerial innovation, and new implements) occurred very late, and did not accelerate markedly until the second half of the 19th century, together with the commercial and financial penetration of rural relationships.

The landholding peasantry had, however, already been granted political citizenship in the first decades of the 19th century, and they were basically confronted with a bourgeois-bureaucratic elite which represented the wider social order, economically and politically.

This relatively independent peasantry were not confronted by a landholding aristocracy, but directly by the tax-collecting state. The peasant uprisings in the late 18th century were spontaneous reactions against public power and bourgeois privileges. Although appealing to the benevolence and justice of the King, they were directed against centralization, deprivation of autonomy, and the mercantilistic assigning of privileges to the commercial bourgeoisie. This reaction was localistic or regionally restricted at first, but later–a couple of decades after the peasantry had been granted political citizenship–it was transformed into a sustained peasant program at the level of national politics. The communalistic commitment was now consistently defended in the parliamentary arena of power.

A notable tendency towards political cleavage within the peasantry –on socio-economic and geographical lines–occurred with increasing differentiation and uneven mercantilization around the middle of the 19th century.

Towards the last quarter of the century the diffusion of neo-technical farming, commodity relationships, and individualized landholding gradually transformed the peasant strategy from localistic defence against *incorporation* in wider networks, towards supra-local defence against the *haphazardness* of the price-regulating market. This move manifested itself in co-operative institutions, and in the exploitation of public services within the framework of a general liberal-democratic movement. The familistic, consumption-based and relatively autonomous character of peasant society had gradually disappeared, under the impact of the dual force of internal differentiation and penetration by commodity relationships. The communalistic orientation of peasant politics had become increasingly irrelevant with the disappearance of a shared, general peasant style of life, although certain elements of the original program effectively survived for a considerable time after the early 1870s as well.

The landed society in *Denmark* was characterized by a highly different agrarian structure, and the sequence of rural transformations was nearly the opposite of the Norwegian one. From the late 18th century to the middle of the 19th century the condition of the peasantry changed from dependent land tenure with client status to market land tenure with citizenship status. The emancipation from above, from the 1780s onwards, implied, firstly, an abolition of personal dependence; secondly, a dissolution of the village clusters and an individualization of landholding; and thirdly, the beginning of a gradual move towards copyhold and peasant proprietorship. The fragmentation of the peasant communities now definitely increased, while the vertical segmentation of landed relationships to some extent prevailed, since the rural aristocracy was still economically and socially dominant. The big landowners were thus the main landed mediators to the national scene during the first half of the 19th century as well, and they retained their control on the local arena.

Contrary to the situation in Norway, agriculture as such was commercially dynamic, with early production for export and for an urban market.

The peasant politics towards the middle of the 19th century were no communalistic defence of peasanthood against intrusion from outside, since commercial landlordism was internal to the agricultural development as such. Peasant politics were at first an elementary struggle for equal status, economic mobility, and individualized proprietorship, in a joint opposition with the pro-liberal and anti-aristocratic bourgeoisie.

The peasant coalition with the crofter class, united by common landed inferiority, later bypassed the urban bourgeoisie and founded a general (peasant-based) liberal-democratic opposition against the urban and rural upper classes. The successful market integration of the peasant proprietors was confirmed with the growing co-operative movement.

The Danish peasantry had thus passed a prolonged 'sack-of-potatoes'-period after the dissolution of communal village life. They developed new types of supra-local associations parallel to the foundation of a political program for an individualized farmer class.

Landed society in *Sweden* was dominated by the privileged nobility, within a social and political framework characterized by legally defined Estates. Different 'types of land' (*jordnatur*) were recognized indices of social rank, although the strict boundaries were gradually obliterated from the 18th century, as the possibilities of peasant land disposal increased. The move towards peasant proprietorship occurred mainly by the purchase of leased crown property to freehold, but also by the dissolution of strictly enforced estate criteria for possession of privileged land: the peasantry (and other non-nobles) were increasingly able to acquire tax-exempted land.

The bulk of the Swedish peasantry was relatively independent–being outside the immediate domain of the landlords–while the village system was regulated jointly by landed interest and public authorities, since there was a notable fusion of officialdom and landlordism at the national level.

The field consolidation of individualized landholding–enclosures and reallotment–was at first enforced on noble lands, while the enclosures on the freeholds occurred notably later. This pattern fits in with the corresponding contrast between the Norwegian and Danish enclosure movement.

Large-scale agriculture was commercially dynamic with production for the foreign market early in the 19th century, and the lowland estates concomitantly led the way in the development of neo-techical farming.

Although the Swedish peasantry had a right of political representation from the early 18th century, this was basically a cooptation within formally strict delineations of status. The peasant participation implied a certain element of power-sharing responsibility, while the range of peasant influence was confined by the non-peasant constellation of forces.

Swedish landed society experienced a peculiar criss-cross of cleavages. There was, as in Norway, a general contrast between the peasant and the non-peasant world. The lord-and-peasant nexus was less intrinsic to the very core of landed relationships, since direct subjugation by powerful landlords was less prevalent than in Denmark. But there was also a crucial antagonism between all non-noble interests and the privileged nobility. This configuration put the peasantry in a sort of stalemate position. They were politically coopted within a general non-noble opposition, and never developed a specific popular movement on their own. Owing to the features of a society based on legally privileged Estates, they had, unlike the Norwegian peasantry, crucial interests in common with the rising bourgeoisie. And owing to their relatively independent position, themselves directly dominating the crofter class, they did not, like the Danish peasantry, share the initial grievances of the mass of rural landless. The individualized peasantry and the rising bourgeoisie jointly opposed the hallmarks of the Estate society, although peasant agriculture later, around 1870, faced the gradually more fused power of upper landed and higher bourgeois interests. The Swedish configuration of rural cleavages thus made specific peasant politics structurally somewhat unlikely–first because the potential antagonism to the bourgeois world was overshadowed by a joint opposition to the privileged landed nobility, and later because the peasantry gradually lost their generic quality as a distinct socio-economic category.

The Estate society in Sweden, with judicial assignment of political rank and authoritatively well-defined principles of social classification, had been obliterated gradually before the formal abolishment of the four-tier *Riksdag* in 1866. But in Norway and Denmark an increasing generalization of monetary criteria as a common denominator also took place in public affairs. A social order that uses money as the basic social measurement implies that a hierarchy of quantitative values replaces the binary opposites which characterize an 'estate society' as well as a simple subsistence economy. This transitional process, of course, started much earlier than the period which is analysed in the present study, but it was, as demonstrated, rather acute from the late 18th century onwards. The change was even formally adopted in Scandinavian suffrage qualifications in the second half of the 19th century.

In chapter 1 we made a conceptual distinction between peasant characteristics which apply to the phenomenon of social *class* and those which apply to the corporate *community*. The 'gemeinschaft' aspects of peasant community mark it off from the wider society as a 'different world', with a specific mode of production. The class character of peasantry stems, on the other hand, from the contradictions within the wider socio-economic system itself. In chapter 5 we then separated these principles of social organization from the legally defined position of peasantry as a social *estate*. From a certain point of view peasantry are adequately described as a particular class or, even more adequately, as several classes. But they may also be an (inferior) estate within a wider judicial-political order. And the local community is definitely crucial to the social organization of peasantry. Thus peasantry have all these qualities, while none of them are exclusively to the point, as argued in Part I. The conclusions which emerge from our comparative and empirical investigation in Parts II–III seem to have considerable theoretical interest:

1) Towards the middle of the 19th century the Scandinavian peasantry increasingly acted as a supra-local social *group*, and thus defied either their corporate localism or their 'sack of potatoes' quality. This (relatively) concerted peasant voice was not, however, primarily based on socio-economic *similarity*. On the contrary: the peasant politics at the national level found expression when the differentiation both within and between the peasant communities had been increasing for several decades.

2) The Scandinavian peasantry engaged in coordinated *class* action precisely by fighting the stigma of structured social inequality: their position as an inferior *estate*. The legal assigning of estate qualities did thus not continue to provide socially stable conditions, but in fact accentuated the political articulation of the peasantry in response to socio-economic change. The particular configuration of 'estate' characteristics was thus (partly) conditional to the transformation of corporate local groups into determined 'class' bodies of action. This transformation implied, moreover, that the problems of the peasantry were generalized beyond the immediate peasant world. The specific 'peasant' characteristics–i.e. the socio-economic peculiarity of the peasant community–were concomitantly obliterated, so peasanthood as such lost its actuality as a mobilizing quality.

3) While the Danish and Swedish peasantry seem to have engaged in coordinated 'class' action parallel to their integration into nationwide socio-economic networks, this was not, however, unconditionally the case in Norway. The Norwegian peasantry constituted itself as a politically conscious 'class' while retaining, and even explicitly by defending–for some time–their basic 'community' characteristics. Conceptually this might seem paradoxical, but it is, as we have argued, intrinsic to the very core of landed relationships in Norway.

The subsequent development indicates, however, a very *genuine* paradox which seems inherent to comprehensive 'peasant politics' regardless of structural contingency:

While the growing 'class quality' of peasantry is conditioned by the dual intrusion of commodity relationships and political 'nationalization', these processes concomitantly split up the peasant communities and reduce the importance of peasant agriculture. Peasant action becomes in itself an agent in this dynamic process, whatever the initial peasant motivation might be. Economically as well as politically peasantry seems to be an intrinsically ambiguous phenomenon.

Bibliography

Alavi, Hamza, "Peasants and Revolution", *The Socialist Register 1965*.

Allardt, Erik, *Samhällsstruktur och sociala spänningar,* Almqvist & Wiksell, Stockholm 1965.

Allardt, Erik and S. Rokkan (eds.) *Mass Politics,* Free Press 1970.

Althusser, Louis, et al., *Lire le Capital,* Maspero, Paris 1967.

Åmark, Karl, *Spannmålshandel och spannmålspolitik i Sverige 1719–1830,* Stockholm 1915.

Amundsen, Kåre, *Norsk sosialøkonomisk historie, 1814–1890,* Universitetsforlaget, Oslo 1963.

Andrén, Nils, *Svensk statskunskap,* Liber, Stockholm 1961.

Andrén, Nils, *Government and Politics in the Nordic Countries,* Almqvist & Wiksell, Stockholm 1964.

Andrén, Nils, et al., *Nordisk statskunskap,* Liber, Stockholm 1963.

Archetti, E., E. Fossum and P. O. Reinton, "Agrarian Structure and Peasant Autonomy", *Journal of Peace Research,* No. 3, 1970.

Aschehoug, T. H., "Statistiske studier over folkemængde og jordbrug i Norges landdistrikter i det 17de og 18de aarhundrede", *Statsøkonomisk Tidsskrift,* Kristiania 1890.

Bacra, Paul, *Utviklingen av den norske jordbruksstatistikk,* Statistisk Sentralbyrå, Oslo 1958.

Bailey, F. G., "The Peasant View of the Bad Life", (1966), in Shanin (ed.) 1971, ch. 22.

Barth, Fredrik, *Models of Social Organization,* London 1966.

Bath, B. H. Slicher van, *The Agrarian History of Western Europe, A. D. 500–1850,* New York 1963.

Beckman, Bjarne, *Dalupproret 1743 och andra samtida rörelser inom allmogen och bondeståndet,* Stockholm 1930.

Belshaw, Cyril, *Traditional Exchange and Modern Markets,* Prentice-Hall 1965.

Benterud, S. J., J. Grude and F. E. Wold, *Meieridriften i Norge indtil aaret 1905,* Kristiania 1908.

Beretninger om Rigets økonimiske tilstand 1827–1855, From the quinquennial reports of the provincial governors (*Amtmennenes Femårsberetninger*). (Norway).

Berg, G. and S. Svensson, *Svensk Bondekultur,* Bonniers, Stockholm 1971.

Bergsgård, Arne, *Ole Gabriel Ueland og bondepolitikken,* vol. I–II, Aschehoug, Oslo 1932.

Bergsgård, Arne, *Norsk historie 1814–1880,* 2.ed., Det norske Samlaget, Oslo 1964.

Bidrag til Bondesamfundets Historie, vol. I–II, Institute for comparative cultural research, Oslo 1933.

Bjerke, Juul, *Langtidslinjer i norsk økonomi 1865–1960,* Samfunnsøkonomiske studier 16, Statistisk Sentralbyrå, Oslo 1966.

Bjørklund, Oddvar, *Marcus Thrane – sosialistleder i et u-land,* 2.ed., Tiden, Oslo 1970.

Bjørkvik, Halvard, *Jord-eige og jordleige i Ryfylke i eldre tid,* Stavanger 1958.

Bjørkvik, Halvard, "The Farm Territories", *Scandinavian Economic History Review,* vol. IV, No. 1, 1956.

Bloch, Marc, *Les caractères originaux de l'histoire rurale francaise,* 2 vols., 2 ed., Paris 1955–56.
Bobé, Louis, G. Graae and Fr. Jürgensen West, *Danske Len,* Copenhagen 1916.
Boethius, Bertil, "New Light on Eighteenth Century Sweden", *Scandinavian Economic History Review,* II, 1954.
Bolin, Sture, "Medieval Agrarian Society in its Prime: Scandinavia", in M. M. Postan (ed.), *The Cambridge Economic History of Europe,* vol. 1, 2.ed., Cambridge University Press 1966.
Borgedal, Paul, *Norges Jordbruk i nyere tid,* vol. I–II, Oslo 1966–68.
Boserup, Ester, *The Conditions of Agricultural Growth,* Allen & Unwin, London 1965.
Brox, Ottar (ed.), *Norsk landbruk – utvikling eller avvikling?,* Pax, Oslo 1969.
Carlsson, Sten, *Ståndssamhälle och ståndspersoner 1700–1865. Studier rörande det svenska ståndssamhällets upplösning,* Gleerup, Lund 1949.
Carlsson, Sten, *Svensk ståndscirkulation 1680–1950,* Uppsala 1950.
Carlsson, Sten & Jerker Rosén, *Svensk Historia,* vol. II, Bonniers, Stockholm 1964.
Chambers, J. D. and G. E. Mingay, *The Agricultural Revolution 1750–1880,* Batsford, London 1966.
Chayanov, A. V., *The Theory of Peasant Economy,* Irwin, Illinois 1966.
Christensen, C., *Agrarhistoriske studier II. Danske landboforhold under enevælden, I,* Copenhagen 1891.
Clout, Hugh, *Agriculture,* Macmillan, London 1971.
Cohn, E., *Privatbanken i København gennem hundrede aar,* Copenhagen 1957.
Coldevin, Axel, *Jordegods og storgårder i Nord-Norge,* Trondheim 1943.
Coldevin, Axel, "Enevoldstiden", *Vårt folks historie,* vol. V, Aschehoug, Oslo 1963.
Dahl, Ottar, "Noen teoretiske problemer i sosialhistorien", *Historisk Tidsskrift,* vol. 37 (1955).
Dahl, Sven, "Strip Fields and Enclosure in Sweden", *Scandinavian Economic History Review,* vol. IX, No. 1, 1961.
Dahrendorf, Ralf, *Class and Class Conflict in Industrial Society,* Routledge & Kegan Paul, London 1959.
Dalton, George, "Peasantries in Anthropology and History", *Current Anthropology,* vol. 13, nos. 3–4, June–Oct. 1972.
Dalton, George (ed.), *Economic Development and Social Change,* Garden City 1971.
Danielsen, Rolf, *Det Norske Storting gjennom 150 år,* vol. II, Gyldendal, Oslo 1964.
Danmarks Historie, vol. 8–11, Politikens Forlag, Copenhagen.
Dannfelt, H. Juhlin, *Kungl. Landtbruksakademien 1813–1912 samt Svenska Landthushållningen under nittonde århundradet,* 2 parts, Stockholm 1913.
Davis, Ioan, *Social Mobility and Political Change,* Macmillan, London 1970.
Den Danske Rigsdag 1849–1949, vol I–VI, Schultz forlag, Copenhagen 1949–53.
Drake, Michael, *Population and Society in Norway 1735–1865,* Cambridge University Press 1969.
Durkheim, E., *Division of Labour in Society,* Free Press, Illinois 1947.
Durkheim, E., *Suicide,* Routledge and Kegan Paul, London 1952.
Dyrvik, Ståle, et al., *Norge under eneveldet 1720–1800,* Universitetsforlaget 1976.
Eisenstadt, S. N., *Modernization – Protest and Change,* Prentice-Hall 1966.
Elgeskog, V., *Svensk torpbebyggelse från 1500-talet till laga skiftet,* L. T.'s förlag, Stockholm 1945.
Elster, Jon, *Nytt perspektiv på økonomisk historie,* Pax, Oslo 1971.
Elster, Jon, *Økonomi og Historie,* Pax, Oslo 1972.
Encyclopedia of the Social Sciences.
Erasmus, C., "Community Development and the Encogido Syndrome", *Human Organization,* vol. 27, no. 1, Spring 1968.
Erslev, Kr., "Europæisk Feudalisme og dansk Lensvæsen", (Danish) *Historisk Tidsskrift,* R. 7, B. 2, Copenhagen 1899–1900.

Fallers, L. A., "Are African Cultivators to be Called 'Peasants'?" *Current Anthropology,* vol. 2, no. 2, April 1961.

Fanon, Franz, *Les damnés de le terre,* Maspero, Paris 1961.

Fedet, Ernest, *The Rape of the Peasantry,* Anchor, New York 1971.

Fink, Troels, *Udskiftningen i Sønderjylland indtil 1770,* Munksgaard, Copenhagen 1941.

Firth, Raymond, *Malay Fishermen: Their Peasant Economy,* 2.ed., London 1966.

Firth, Raymond, *Primitive Polynesian Economy,* London 1939.

Firth, Raymond and Yamey (eds.), *Capital, Saving and Credit in Peasant Societies,* Allen & Unwin, London 1964.

Fladby, Rolf, "Bønder og embetsmenn i lokalstyringen etter 1837", *Historisk Tidsskrift,* vol. 46 (1967).

Forman, S. and J. F. Riegelhaupt, "Market Place and Marketing Systems: Toward a Theory of Peasant Economic Integration", *Comparative Studies in Society and History,* vol. 12, no. 2, (April) 1970.

Foster, G. M., "What Is Folk Culture?", *American Anthropologist,* LV, no. 2, (April–June 1953).

Framveksten av de politiske partier i de nordiske land på 1800-tallet, Problemer i nordisk historieforskning II, Universitetsforlaget, Bergen 1964.

Franklin, S. H., *European Peasantry: The final phase,* Methuen, London 1969.

Franklin, S. H., *Rural Societies,* Macmillan, London 1971.

Franklin, S. H., "Reflections on the Peasantry", *Pacific Viewpoint,* vol. 3, no. 1, March 1962.

Fridericia, J. A., "Historisk-statistiske Undersøgelser over Danmarks Landboforhold i det 17de Aarhundrede", (Danish) *Historisk Tidsskrift,* R. 6, B. 2, 1889–90.

Fridlizius, Gunnar, "Sweden's Exports 1850–1960", *Economy and History,* vol. VI, 1963.

Friis, Jacob, *Marcus Thrane,* Kristiania 1917.

Frimannslund, Rigmor, "Farm Community and Neighborhood Community", *Scandinavian Economic History Review,* vol. IV, no. 1, 1956.

Galeski, B., *Basic Concepts of Rural Sociology,* Manchester University Press 1972.

Geertz, Clifford, "Studies in Peasant Life: Community and Society", *Biennial Review of Anthropology 1961.*

Gellerman, Olle, *Staten och jordbuket 1867–1918,* Almqvist & Wiksell, Stockholm 1958.

Gellner, Ernest, *Thought and Change,* Weidenfeld and Nicolson, London 1964.

Gellner, Ernest, *Saints of the Atlas,* Weidenfeld and Nicolson, London 1969.

Gellner, Ernest, "The Great Patron. A reinterpretation of tribal rebellions", *Archives Européennes de Sociologie,* vol. X, no. 1, 1969.

Genovese, Eugene, *The Political Economy of Slavery,* Random House, New York 1965.

Godelier, Maurice, *Rationalité et irrationalité en économie,* Maspero, Paris 1966.

Goodwin, A. (ed.), *The European Nobility in the Eighteenth Century,* Black, London 1953.

Grankvist, Rolf, *Thranitterbevegelsen i Trøndelag,* Trondheim 1966.

Grendahl, T. (ed.), *Jordskifteverket gjennom 100 år. 1859–1958,* Oslo 1959.

Grude, Egil Harald, *Skatt, bønder og embetsmenn. Ekstraskatten i Stavanger amt 1762–65,* unpublished thesis in history, University of Oslo 1972.

Gurr, T. R., *Why Men Rebel,* Princeton University Press 1970.

Haarstad, Kjell, *Bondenæringen i støpeskjeen.* Universitetsforlaget, Oslo 1976.

Hals, Anne, *Sparebanken og det store hamskiftet. En analyse av Gran Sparebanks utlånspolitikk 1857–1887,* unpublished thesis in history, University of Oslo 1971.

Hammond, J. L. and Barbara, *The Village Labourer,* London 1912.

Hansen, K. (ed.), *Det danske Landbrugs historie,* vol. I–V, Gads forlag, Copenhagen 1934–45.

Hansen, Svend Aage, *Adelsvældens grundlag,* Gads forlag, Copenhagen 1964.

Hansen, Svend Aage, *Pengevæsen og kredit 1813–1860,* Odense 1960.
Hansen, Svend Aage, *Økonomisk vækst i Danmark,* vol. I–II, Copenhagen 1972–73.
Hasund, Sigvald, *Vårt landbruks historie,* Aschehoug, Oslo 1932.
Hasund, Sigvald, *Det kgl. Selskap for Norges Vel,* II, Gjøvik 1941.
Hasund, Sigvald, "Grendskipnad og utskiftning", *Syn og Segn,* vol. XX, 1914.
Hasund, Sigvald, *Bønder og stat under naturalsystemet,* Olaf Norlis forlag, Kristiania 1924.
Heckscher, Eli, *Sveriges ekonomiska historia från Gustav Vasa,* vol. I–II, Bonniers, Stockholm 1949.
Heckscher, Eli, *Svenskt arbete och liv,* Bonniers, Stockholm 1957.
Herlitz, Lars, *Jordegendom och ränta,* Lund 1974.
Herlitz, Nils, *Grunddragen i det svenska statsskickets historia,* Svenska bokförlaget, 6. ed., Stockholm 1964.
Hicks, John, *A Theory of Economic History,* Oxford 1969.
Historical Statistics 1968, Statistisk Sentralbyrå, Oslo.
Historisk statistik för Sverige, I–III, Statistisk Centralbyrån, Stockholm 1955–60.
Hobsbawn, Eric, *Primitive Rebels,* Norton, New York 1959.
Hobsbawn, Eric, "Peasants and Rural Migrants in Politics", in C. Veliz (ed.), *The Poltics of Conformity in Latin America,* Oxford University Press 1967.
Hobsbawn, Eric, "Peasants and Politics", *The Journal of Peasant Studies,* vol. 1, no. 1 (Oct, 1973).
Hobsbawn, Eric and G. Rudé, *Captain Swing,* Lawrence & Wishart, London 1969.
Høgh, Erik, *Vælgeradfærd i Danmark 1849–1901,* Copenhagen 1972.
Holm, Edward, *Kampen om Landboreformene i Danmark i Slutningen af 18. Aarhundrede,* Copenhagen 1888.
Holmsen, Andreas, *Norges Historie,* 3. ed., Universitetsforlaget, Oslo 1961.
Holmsen, Andreas, *Gard, Bygd, Rike,* Universitetsforlaget, Oslo 1966.
Holmsen, Andreas, "From Tenancy to Freehold Peasant Ownership", *Scandinavian Economic History Review,* IX, 1961.
Holmsen, Andreas, "The Old Norwegian Peasant Community", *Scandinavian Economic History Review,* IV, no. 1, 1956.
Holmsen, Andreas, "Landowners and Tenants in Norway", *Scandinavian Economic History Review,* VI, 1958.
Holmsen, Andreas, "Eldre norsk tunskipnad", *Heimen,* vol. XIV, June 1969.
Holmsen, Andreas and H. Bjørkvik, "Kven åtte jorda i den gamle leilendingtida?", *Heimen,* vol. IX (1952–54).
Hovde, B. J., *The Scandinavian Countries 1720–1865,* 2 vols., Chapman & Grimes, Boston 1943.
Huizer, G., "'Resistance to Change' and Radical Peasant Mobilization: Foster and Erasmus Reconsidered", *Human Organization,* vol. 29, no. 4, Winter 1970.
Huizer, G., "Land Invasion as a Non-Violent Strategy of Peasant Rebellion", *Journal of Peace Research,* No. 2, 1972.
Huizer, G., *The Revolutionary Potential of Peasants in Latin America,* Lexington, Mass. 1972.
Hundre års historisk forskning, Universitetsforlaget, Oslo 1970.
Hvidt, Kristian, *Flugten til Amerika,* Copenhagen 1971.
Hvidt, Kristian, (ed.), *Emigrationen fra Norden indtil 1. Verdenskrig,* Copenhagen 1971.
Ingers, Enoch, *Bonden i svensk historia,* 3. vols., Stockholm 1943–56.
Ionescu, G. and E. Gellner (eds.), *Populism,* Weidenfeld and Nicolson, London 1969.
Jacobsen, Knut Dahl, *Teknisk hjelp og politisk struktur,* Universitetsforlaget, Oslo 1964.
Jansen, Svein, *Bondevennbevegelsen i Norge 1865–1880,* unpublished thesis in history, University af Bergen 1965.
Jensen, Adolph, "Befolkningsforhold i de nordiske lande i det 18. Aarhundrede", *Nationaløkonomisk Tidsskrift,* Copenhagen 1935.

Jensen, A., "De høje Priser paa Landejendomme", *Nationaløkonomisk Tidsskrift,* 1912.
Jensen, Hans, *Dansk Jordpolitik 1757–1919,* vol. I, Nordisk Forlag, Copenhagen 1936.
Jensen, Hans, *De Danske Stænderforsamlingers Historie,* vol. I–II, Copenhagen 1931–34.
Jensen, Hans, "Bondecirkulæret af 8. Novmber 1845. Et Bidrag til Kristian 8.s Tids Historie", (Danish) *Historisk Tidsskrift,* 9. R. II, Copenhagen 1921–23.
Jensen, Sigurd, *Fra patriarkalisme til pengeøkonomi,* Gyldendal, Copenhagen 1950.
Jörberg, Lennart, *The Industrial Revolution in Scandinavia 1850–1914,* Fontana 1970.
Jörberg, Lennart, "Structural Change and Economic Growth: Sweden in the 19th Century", *Economy and History,* vol. VIII, 1965.
Jørgensen, A., "Borger og bonde på Møn i 18. århundrede", *Arkiv,* vol. I (1966).
Johnsen, Oscar Albert, *Norges Bønder,* Aschehoug, Kristiania 1919.
Johnsen, Oscar Albert, "Fra leilending til selveier", *Nordisk Tidsskrift* 1910.
Jutikkala, Eino, *Bonden i Finland genom tiderna,* LT's Förlag, Helsinki 1963.
Kaartvedt, Alf, *Kampen mot parlamentarisme 1880–1884,* Universitetsforlaget, Oslo 1956.
Kaartvedt, Alf, *Det Norske Storting gjennom 150 år,* vol. I, Gyldendal, Oslo 1964.
Kaartvedt, Alf and Leif Chr. Hartsang, *Kongeriket Norges Hypotekbank 1852–1952,* Oslo 1952.
Karlbom, R., *Bakgrunden till 1809 års regeringsform,* Gothenburg 1964.
Keilhau, W., *Det Norske Folks Liv og Historie* (abb. NFLH), vols. VIII–X.
Klokk, O., *Det norske landbruk siden omkring 1750,* Kristiania 1920.
Koht, Halvdan, "Prisar og politikk i norsk historie", *Samtiden* 1913.
Koht, Halvdan, *Thranitterbevægelsen,* Kristiania 1917.
Koht, Halvdan, *Norsk bondereising,* Aschehoug, Oslo 1926.
Koht, Halvdan, *Johan Sverdrup,* vol. I–II, Aschehoug, Oslo 1918–22.
Koht, Halvdan, *Nye innhogg og utsyn,* Aschehoug, Oslo 1964.
Koht, Halvdan (ed.), *Fra den gamle bondeopposisjon. Brev og innlegg fra Stortingene 1818, 1821, 1822 og 1833 av Jacob Hoel,* Historisk Tidsskrift 2.–4. levering 1926, Oslo 1927.
Kornhauser, William, *The Politics of Mass Society,* Routledge of Kegan Paul, London 1959.
Kristiansen, O., *Samferdsel i Norge 1814–1830,* Oslo 1926.
Kroeber, A. L., *Anthropology,* Harcourt, New York 1948.
Krokann, Inge, "Det store hamskiftet i bondesamfunnet", *Norsk kulturhistorie,* vol. V, Cappelen, Oslo 1942.
Kuhnle, Stein, "Stemmeretten i 1814", *Historisk Tidsskrift,* nr. 4/1972.
Kulturhistorisk Leksikon for nordisk middelalder.
Kumm, Elfred, *Jordhunger och dagsverkstorp,* Kooperativa Förbundets Bokförlag, Stockholm 1949.
Kuuse, Jan, *Från redskap till maskiner. Mekaniseringsspridning och kommersialisering inom svensk jordbruk 1860–1910,* Gothenburg 1970.
Kuuse, Jan, "Mechanisation, Commercialisation and the Protectionist Movement in Swedish Agriculture 1860–1910", *Scandinavian Economic History Review,* vol. XIX, no. 1, 1971.
Laclau, E., "Feudalism and Capitalism in Latin America", *New Left Review,* no. 67, 1971.
Lafferty, W. M., *Economic Development and the Response of Labor in Scandivavia,* Universitetsforlaget, Oslo 1971.
Landsberger, Henry (ed.), *Latin American Peasant Movements,* Cornell University Press 1969.
Landsberger, Henry (ed.), *Rural Protest: Peasant Movements and Social Change,* Macmillan, London 1974.

Laslett, Peter, *The World We Have Lost*, 2.ed., Methuen, London 1971.

Lassen, Aksel, *Fald og Fremgang. Træk af befolkningsudviklingen i Danmark 1645–1960*, Universitetsforlaget i Aarhus 1965.

Lassen, Aksel, "The Population of Denmark, 1660–1960", *Scandinavian Economic History Review*, XIV, 1966.

LeClair, E. E. and H. K. Schneider (eds.), *Economic Anthropology*, New York 1968.

Lefebvre, Henri, *Du rural à l'urbain*, ed. anthropos, Paris 1970.

Lieberman, Sima, *The Industrialization of Norway 1800–1920*, Universitetsforlaget, Oslo 1970.

Lindhardt, P. G., *Vækkelser og kirkelige retninger i Danmark*, Det Danske Forlag, Copenhagen 1951.

Lorenz, Einhard, *Arbeiderbevegelsens historie 1789–1930*, Pax, Oslo 1972.

Lorenz, Einhard, (ed.), *Arbeidere, forén dere!*, Pax, Oslo 1969.

Lund, Carl, *Jaabæk og bondevennbevegelsen i Bratsberg amt*, (handwritten) thesis in history, University of Oslo 1929.

Lunden, Kåre, *Økonomi og Samfunn*, Universitetsforlaget, Oslo 1972.

Lunden, Kåre, "Agrartilhøve ca. 1600–1720", *Historisk Tidsskrift*, vol. 48, no. 4, 1969.

Lunden, Kåre, "Om årsakene til den norske bondefridomen", *Heimen*, XV, 1971.

Lunden, Kåre, "Framveksten av samvirketiltak mellom norske mjølkeprodusenter i tida 1856–1905", stencil, Oslo 1970.

Martinius, Sture, *Agrar kapitalbildning och finansiering 1833–1892*, Gothenburg 1970.

Martinius, Sture, *Jordbruk och ekonomisk tillväxt i Sverige 1830–1870*, Gothenburg 1970.

Marx, Karl, *Pre-capitalist Economic Formations*, Lawrence & Wishart, London 1964.

Marx, Karl, *The 18th Brumaire of Louis Bonaparte*, – (publ. 1852).

Meillassoux, Claude, "From Reproduction to Production", *Economy and Society*, vol. 1, no. 1, 1972.

Meillassoux, Claude, "Essai d'interprétation du phénomène économiqe dans les sociétés traditionelles d'autosubsistence", *Cahiers d'études africaines*, no. 4, 1960.

Meyer, Poul, *Danske Bylag*, Copenhegen 1949.

Miller, Kenneth E., *Government and Politics in Denmark*, Houghton Mifflin, Boston 1968.

Millward, Roy, *Scandinavian Lands*, Macmillan, London 1969.

Mingay, G. E., *Enclosure and the Small Farmer in the Age of the Industrial Revolution*, Macmillan, London 1968.

Moe, Thorvald, *Demographic Developments and Economic Growth in Norway, 1740–1940*, Ann Arbor, Michigan 1970.

Mohn, J. N., "Storthingene 1815–1874, belyste ved Statistik", *Norsk Retstidende*, 1874.

Montgomery, Arthur, *Industrialismens genombrott i Sverige*, Skoglunds Bokförlag, Stockholm 1931.

Moore, Barrington, Jr., *Social Origins of Dictatorship and Democracy. Lord and peasant in the making of the modern world*, Beacon Press 1967.

Mortensson, Ivar, *Bondeskipnad i Norig i eldre tid*, Olaf Norli, Kristiania 1904.

Mousnier, Roland, *Peasant Uprisings in the 17th Century France, Russia and China*, Allen & Unwin, London 1971.

Munch, Peter A., "The Peasant Movement in the Scandinavian Countiries", in *Democratic Folk Movements in Scandivavia*, Minneapolis 1951.

Munch, Peter A., "The Peasant Movement in Norway: A Study in Class and Culture", *The British Journal of Sociology*, vol. V, no. 1, 1954.

Mykland, Knut, *Skiftet i forvaltningsordningen i Danmark og Norge i tiden fra omkring 1630 og inntil Frederik den tredjes død*, Universitetsforlaget, Bergen 1973.

Mykland, Knut, (ed.), *Omkring 1814*, Gyldendal, Oslo 1967.

Mykland, Knut et al., *Norge under eneveldet 1660–1720*, Universitetsforlaget 1975.

Nash, Manning, *Primitive and Peasant Economic Systems,* Chandler, San Francisco 1966.
Nerbøvik, Jostein, *Norsk Historie 1870–1905,* Det norske Samlaget, Oslo 1973.
Norsk Biografisk Leksikon.
Norsk Forfatter-Lexikon.
Nygren, Ingemar, *Svensk sparebanksutlåning. 1820–1913,* Gothenburg 1967.
Obregón, A. Q., "Contemporary Peasant Movements", in Lipset & Solari (eds.), *Elites in Latin America,* Oxford University Press 1967.
Østberg, Kristian, *Norsk Bonderett,* vol. XI, Oslo 1936.
Østerud, Øyvind, "Nytt perspektiv på det store hamskifte", (Norwegian) *Historisk tidsskrift,* No. 2, 1975.
Østerud, Øyvind, "The Class Quality of Peasantry: a conceptual exploration", *Scandinavian Political Studies,* vol. 11 (1976).
Østerud, Øyvind, "The Transformation of Scandinavian Agrarianism", *Scandinavian Journal of History,* 1 (1976).
Østerud, Øyvind, "Configurations of Scandinavian Absolutism", paper to ISSC/MSH Symposium, Bellagio, October 1976.
Øverland, O. A., *Thraniterbevægelsen, Kristiania 1903.*
Olsen, Erling, *Danmarks økonomiske historie siden 1750,* Copenhagen 1961.
Olsen, Gunnar, *Hovedgård og bondegård. Studier over stordriftens udvikling i Damnark i tiden 1525–1774,* Landbohistoriske Skrifter, vol. 1, Rosenkilde & Bagger, Copenhagen 1957.
Olsen, Gunnar, *Træhesten, hundehullet og den spanske kappe,* Copenhagen 1960.
Olsson, Gunnar, *Hattar och mössor. Studier över partiväsendet i Sverige 1751–1762,* Akademiförlaget, Gothenburg 1963.
Olsson, Ragnar, *Bondeståndet under den tidigare frihetstiden. Val, organisation, och arbetssätt,* Lund 1926.
Olsson, Ragnar, *Riksdagsmannavalen till bondeståendet under den senare delen av frihetstiden (1740–1772),* Lund 1948.
Pizzorno, A. (ed.), *Political Sociology,* Penguin 1971.
Polanyi, Karl, *The Great Transformation,* New York 1944.
Polanyi, Karl, *Primitive, Archaic and Modern Economies* (ed. by George Dalton), Doubleday 1968.
Polanyi, Karl et al., *Trade and Market in the Early Empires,* Free Press 1957.
Potter, J. M., M. M. Diaz and G. M. Foster (eds.), *Peasant Societies: A Reader,* Little, Brown, 1967.
Rasmussen, Erik, *Komparativ Politik,* 2 vols., Gyldendal, Copenhagen 1969.
Redfield, Robert, *The Primitive World and Its Transformations,* Cornell University Press 1953.
Redfield, Robert, *Peasant Society and Culture,* University of Chicago 1956.
Reinton, Lars, *Seterbruket i Noreg,* vol. I–III, Oslo 1955–61.
Reinton, Lars, *Folk og fortid i Hol,* vol. II, Oslo 1943.
Rise Hansen, C. and Axel Steensberg, *Jordfordeling og udskiftning. Undersøgelser i tre sjællandske landsbyer,* Mungsgaard, Copenhagen 1951.
Robberstad, Knut, *Fyrelesningar um Rettssoga i millomalder og ny tid,* vol. II, 4. ed., Universitetsforlaget, Oslo 1966.
Rockwell, Joan, "The Danish Peasant Village", *The Journal of Peasant Studies,* Vol. 1, no. 4 (July 1974).
Rokkan, Stein, *Citizens, Elections, Parties,* Universitetsforlaget, Oslo 1970.
Rokkan, Stein, & Jean Meyriat (eds.), *International Guide to Electoral Statistics,* Vol. I: National Elections in Western Europe, Mouton, Paris 1969.
Rokkan, Stein and R. Bendix, "The Extension of Citizenship to the Lower Classes", in Mattei Dogan & Richard Rose (eds.), *European Politics,* Little, Brown, Boston 1971.

(S. M. Lipset and) Stein Rokkan, (eds.), *Party Systems and Voter Alignments,* The Free Press, New York 1967.

Ross, Alf, *Dansk Statsforfatningsret,* 2. rev. ed., Nyt Nordisk Forlag, Copenhagen 1966.

Rudé, George, *The Crowd in History,* Wiley and Sons, 1964.

Runblom, Harald, and Hans Norman (eds.), *From Sweden to America. A History of the Migration,* University of Minnesota Press, Minneapolis 1976.

Rygg, Nicolai, *Norges Banks historie,* vol. I, Kristiania 1918.

Sahlins, Marshall, *Stone Age Economics,* Tavistock, London 1974.

Sandvik, Gudmund, "Jordleigar og jordeigar i norsk historie", stencil, University of Oslo 1969.

Sandvik, Gudmund, *Det gamle veldet,* Gyldendal, Oslo 1975.

Sars, Ernst, *Samlede Værker,* Kristiania and Copenhagen 1912.

Schweigaard, A. M., *Norges Statistik,* Christiania 1840.

Seip, Anne-Lise, "Tokammerhøyre foran forfatningskampen i 1870-årene", *Historisk Tidsskrift,* no. 2, 1973.

Seip, Jens Arup, *Et regime foran undergangen,* Gyldendal, Oslo 1945.

Seip, Jens Arup, *Teorien om det opinionsstyrte enevelde,* Universitetsforlaget, Oslo 1958.

Seip, Jens Arup, *Tanke og handling i norsk historie,* Gyldendal, Oslo 1968.

Seip, Jens Arup, *Ole Jacob Broch og hans samtid,* Gyldendal, Oslo 1971.

Seip, Jens Arup, *Utsikt over Norges historie,* I, Gyldendal, Oslo 1974.

Sejersted, Francis, *Norges Bank og høykonjunkturen i 1840-årene,* Universitetsforlaget, Oslo 1968.

Sejersted, Francis, *En teori om den økonomiske utvikling i Norge i det 19. århundre,* stencil, University of Oslo 1973.

Semmingsen, Ingrid, *Veien mot Vest,* vol. I–II, Oslo 1941–50.

Semmingsen, Ingrid, "The Dissolution of Estate Society in Norway", *Scandinavian Economic History Review,* vol. II, no. 2, 1954.

Semmingsen, Ingrid, "Ludvig Daae og de politiske gruppedannelser i 1870-årene", *Historisk Tidsskrift,* vol. 36 (1952–53).

Shanin, Teodor (ed.), *Peasants and Peasant Societies,* Penguin 1971.

Shanin, Teodor, *The Awkward Class,* Clarendon Press, Oxford 1971.

Shanin, Teodor, "Peasantry: Delineation of a Sociological Concept and a Field of Study", *Archives Européennes de Sociologie,* tome XII, no. 2, 1971.

Shanin, Teodor, "The Nature and Logic of the Peasant Economy", *The Journal of Peasant Studies,* vol. 1, no. 1 (Oct. 1973) and no. 2 (Jan. 1974).

Siegel, B. J. (ed.), *Biennial Review of Anthropology,* Stanford.

Skappel, Simen, *Om husmannsvesenet i Norge,* Kristiania 1922.

Skappel, Simen, *Eiendomsbesiddere i Norges Landdistrikter, Kristiania 1918.*

Skappel, Simen, *"De høie priser paa jordeiendomme",* Statsøkonomisk Tidsskrift, 1916.

Skappel, Simen, "Det norske jordleievesens geografiske utbredelse i eldre tid og overgangen fra leie til selveie", *Historisk Tidsskrift,* vol. 30, Oslo 1934–36.

Skrubbeltrang, Fridlev, *Den Danske Bonde 1788–1938,* Copenhagen 1938.

Skrubbeltrang, Fridlev, *Den Danske Husmand,* vol. I–II, Det Danske Forlag, Copenhagen 1952–54.

Skrubbeltrang, Fridlev, "Developments in Tenancy in Eighteenth Century Denmark as a Move towards Peasant Proprietorship", *Scandinavian Economic History Review,* IX, 1961.

Skullerud, Aage, *Bondeopposisjonen og religionsfriheten i 1840-årene,* Universitetsforlaget, Oslo 1971.

Slettan, Dagfinn, *Søren Jaabæk og bondevennbevegelsen i Lister og Mandals amt,* Universitetsforlaget, Oslo 1974.

Smedberg, Staffan, *Frälsebonderörelser i Halland och Skåne 1772–76,* Läromedelsförlagen, Uppsala 1972.

Smelser, Neil, *Theory of Collective Behaviour,* Routledge & Kegan Paul, London 1962.
Smith, A. D., *The Concept of Social Change,* Routledge & Kegan Paul, London 1973.
Smitt, Jonas, *Det Norske Landbrugs Historie,* Kristiania 1875.
Smitt, Jonas, "Om grundeiendommens udtrykning og dennes betydning for vort jordbrug", *Statsøkonomisk Tidsskrift* Kristiania 1890.
Sommarin, Emil, *Vårt sparbankväsen 1834–1892,* Lund 1942.
Statistiske Oversikter 1926, Statistisk Sentralbyrå, Oslo.
Stavenhagen, Rodolfo (ed.), *Agrarian Problems and Peasant Movements in Latin America,* Anchor Books 1970.
Steen, Sverre, *Det Norske Folks Liv og Historie* (abb. NFLH), vol. IV–VII.
Steen, Sverre, *Det frie Norge,* vol. I–V, Cappelen, Oslo 1952–62.
Steen, Sverre, *Tusen års norsk historie,* Cappelen, Oslo 1958.
Steen, Sverre, *Lokalt selvstyre i Norges bygder,* Cappelen, Oslo 1968.
Steen, Sverre, *Amt og stat 1837–1860,* Cappelen, Oslo 1973.
Steward, Julien, "Area Research, Theory and Practice", *Social Science Research Bulletin,* No. 63, New York 1950.
Storthings-Efterretninger 1836–54.
Studier i sparing og sparebankvesen i Norge 1822–1972, Gyldendal, Oslo 1972.
Sundbärg, Gustav, *Bevölkerungsstatistik Schwedens 1750–1900,* Stockholm 1907.
Svalestuen, Andres, *Timms emigrasjonshistorie,* Oslo 1973.
Svalestuen, Andres, (Comparative outline on Nordic emigration in Hvidt (ed.) 1971).
Sverdrup, Georg, *Lofthusbevægelsen,* Kristiania 1917.
Tank, G., "Salg av statsgods efter 1814", *Historisk Tidsskrift,* 5. R. 6. B., Oslo 1927.
Terrey, Emmanuel, *Le matérialisme historique devant les sociétés segmentaires et lignagère,* Maspero, Paris 1969.
Thermænius, Edvard, *Lantmannapartiet. Dess uppkomst, organisation och tidigare utveckling,* Uppsala 1928.
Thermænius, Edvard, (ed.), *Svensk Bondepolitik,* Gebers förlag, Stockholm 1931.
Thomson, A., *Grundskatterna i den politiska diskussionen 1809–1866,* Lund 1923.
Try, Hans, *Gardsskipnad og og bondenæring,* Universitetsforlaget, Oslo 1969.
Tveite, Stein, *Jord og gjerning,* Bøndenes forlag, Oslo 1959.
Tveite, Stein, *Bønder og prisskifte,* unpublished thesis in history, University of Oslo 1953.
Utterstrøm, Gustaf, "Population Problems in Pre-industrial Sweden", *Scandinavian Economic History Review,* vol. 2, no. 2, 1954.
Utterstrøm, Gustaf, "Population and Agriculture in Sweden ca. 1700–1830", *Scandinavian Economic History Review,* vol. IX, 1961.
Utterstrøm, Gustaf, *Jordbrukets arbetare,* 2 vols., Tidens förlag, Stockholm 1957.
Valen-Sendstad, Fartein, *Norske landbruksredskaper. 1800–1850-årene,* Lillehammer 1964.
Valgmandsvalgene og Stortingsvalgene 1815–1885, NOS III, nr. 219.
Verney, Douglas, *Parliamentary Reform in Sweden 1866–1921,* Clarendon Press, Oxford 1957.
Visted, K. and H. Stigum, *Vår gamle bondekultur,* vol. I, Cappelen, Oslo 1951.
Wallin, Gunnar, *Valrörelser och valresultat. Andrakammarvalen i Sverige 1866–1884,* Christophers, Stockholm 1961.
Weber, Max, *Wirtschaft und Gesellschaft,* 4th ed., Mohr, Tübingen 1956.
Wilkinson, Paul, *Social Movement,* Macmillan, London 1971.
Winberg, Christer, *Folkökning och proletarisering,* University of Gothenburg 1975.
Winding, Kjeld, *Den Danske Bondebevegelse,* Det Danske Selskab, Copenhagen 1943.
Wohlin, Nils, *Den jordbruksidkande befolkningen i Sverige 1751–1900,* Emigrationsutredningen, bil. 9, Stockholm 1909.
Wohlin, Nils, *Den svenska jordstyckningspolitiken i 18:e och 19:e århundradena,* Stockholm 1912.

Wolf, Eric, *Peasants,* Prentice-Hall 1966.

Wolf, Eric, *Peasant Wars of the Twentieth Century,* Harper, New York 1969.

Wolf, Eric, "Types of Latin American Peasantry: A Preliminary Discussion", *American Anthropologist,* LVII, no. 3, 1955.

Wolf, Eric, "On Peasant Pebellions", *International Social Science Journal,* vol. XXI, no. 2, 1969.

Woloch, Isser (ed.), *The Peasantry in the Old Regime, Conditions and Protests,* Holt, Rinehart and Winston 1970.

Worsley, Peter, *The Third World,* Weidenfeld and Nicolson, London 1967.

Worsley, Peter, (ed.), *Two Blades of Grass,* Manchester University Press 1971.

Index of Proper Names